1

Taking Off Quantities:
Civil Engineering

Other titles from E & FN Spon

Building Regulations Explained
1992 Revision
J. Stephenson

CESMM3 Explained
B. Spain

Commercial Estimator
Marshall & Swift

Residential Estimator
Marshall & Swift

Construction Contracts
Law and Management
J. Murdoch and W. Hughes

Construction Tendering and Estimating
J. I. W. Bentley

Effective Speaking
Communicating in speech
C. Turk

Effective Writing
Improving scientific, technical and business communication
C. Turk and J. Kirkman

Estimating Checklists for Capital Projects
2nd edition
The Association of Cost Engineers

Good Style
Writing for science and technology
J. Kirkman

Housing Defects Reference Manual
The Building Research Establishment Defect Action Sheets
Building Research Establishment

Post-Construction Liability and Insurance
Edited by J. Knocke

The Presentation and Settlement of Contractors' Claims
G. Trickey

Project Budgeting for Buildings
D. Parker and A. Dell'Isola

Project Management Demystified
Today's tools and techniques
G. Reiss

Risk Analysis in Project Management
J. Raftery

Spon's Budget Estimating Handbook
2nd edition
Spain and Partners

Spon's Building Costs Guide for Educational Premises
Tweeds

Spon's Construction Cost and Prices Indices Handbook
M. Flemming and B. Tysoe

Spon's Asia Pacific Construction Costs Handbook
Davis Langdon & Everest

Spon's European Construction Costs Handbook
Davis Langdon & Everest

Standard Method of Specifying of Minor Works
3rd edition
L. Gardiner

Understanding JCT Building Contracts
4th edition
D.M. Chappell

Value Management in Design and Constructions
S. Male and J. Kelly

Write in Style
A guide to good English
R. Palmer

Spon's Contractors' Handbook Series
Prices, costs and estimates for all types of building works from £50 to £50,000 in value.
Minor Works, Alterations, Repairs and Maintenance
Plumbing and Domestic Heating
Painting, Decorating and Glazing
Electrical Installation

Spon's Price Book Series
For building works over £50,000 in value, covering all types and sizes of construction contracts.
Spon's Architects' and Builders' Price Book
Spon's Civil Engineering and Highway Works Price Book
Spon's Landscape and External Works Price Book
Spon's Mechanical and Electrical Services Price Book

Taking Off Quantities:
Civil Engineering

Edited by Tweeds

E & FN SPON
An Imprint of Chapman & Hall
London · Glasgow · Weinheim · New York · Tokyo · Melbourne · Madras

Published by E & FN Spon, an imprint of Chapman & Hall,
2–6 Boundary Row, London SE1 8HN, UK

Chapman & Hall, 2–6 Boundary Row, London SE1 8HN, UK

Blackie Academic & Professional, Wester Cleddens Road, Bishopbriggs, Glasgow G64 2NZ, UK

Chapman & Hall GmbH, Pappelallee 3, 69469 Weinheim, Germany

Chapman & Hall USA, One Penn Plaza, 41st Floor, New York, NY 10119, USA

Chapman & Hall Japan, ITP-Japan, Kyowa Building, 3F, 2-2-1 Hirakawacho, Chiyoda-ku, Tokyo 102, Japan

Chapman & Hall Australia, Thomas Nelson Australia, 102 Dodds Street, South Melbourne, Victoria 3205, Australia

Chapman & Hall India, R. Seshadri, 32 Second Main Road, CIT East, Madras 600 035, India

First edition 1995

Printed in Great Britain by T.J. Press (Padstow) Ltd, Padstow, Cornwall.

ISBN 0 419 20400 8

A catalogue record for this book is available from the British Library

♾ Printed on permanent acid-free paper, manufactured in accordance with ANSI/NISO Z 39.48-1992 and ANSI/NISO Z 39.48-1984 (Permanence of Paper).

CONTENTS

PREFACE

The role of the quantity surveyor is changing rapidly and he is now expected to provide project and financial management services in addition to his traditional expertise. But whatever new skills are acquired, he must still possess a sound knowledge of building construction and the ability to take-off quantities from drawings.

CESMM 3 Explained was published in 1992 and was described as the definitive work on civil engineering measurement. Since publication, discussions have taken place with quantity surveyors, engineers, academics and students and it appeared that there was a need for a book containing examples of civil engineering taking-off only.

This book, *Taking-Off Quantities - Civil Engineering*, re-presents the appendices from *CESMM 3 Explained* together with the first two chapters which deal with general principles of measurement and how CESMM 3 works. Although it is expected that civil engineering and quantity surveying students will form the major part of the readership, interest has already been expressed by practising engineers and surveyors on the need for a book providing examples of civil engineering taking-off accompanied by a commentary on the measurement techniques being used.

Despite the reduction in the number of disputes since Dr Martin Barnes produced CESMM 1 in 1976, disagreements over the definitive way to measure engineering work continue. It is hoped that this book can play a part in reducing this number even further and also save time and money in expensive litigation and arbitration proceedings.

I am indebted to Rona Harper, Neil Harper and Nikki Lark for their calligraphic skills, Paul Spain for presentation and Gil Nicholls who prepared the drawings. I am also grateful to Stephen Booth and the Institution of Civil Engineering Surveyors for permission to reproduce some of the information in Chapter 9. Finally, I would particularly like to thank Len Morley for the major role he played in the preparation of the taking-off examples.

I would welcome constructive criticism of the book together with suggestions for improving its scope and contents. Whilst every effort has been made to ensure the accuracy of the information given in this publication, neither the author nor the publishers accept liability in any way or of any kind resulting from the use made by any person of such information.

There are now many women working in the construction industry; where the pronoun 'he' is used it applies to both men and women.

Bryan J.D. Spain, FInstCES, MACostE
TWEEDS
Chartered Quantity Surveyors
Cavern Walks
8 Matthew Street
Liverpool L2 6RE

Chapter 1
GENERAL PRINCIPLES OF MEASUREMENT

Until comparatively recently, the person preparing the Bill of Quantities - the 'taker-off' - had a limited choice of how to convert the information on the drawings into a Bill of Quantities.

Traditionally the systems followed a procedure of:

Taking off - measuring from the drawings and entering the dimensions on to specially ruled dimension paper

Squaring - calculating and totalling the lengths, areas and volumes of the dimensions

Abstracting - collecting the totals from the dimension paper on to an abstract to produce a final total for each individual description

Billing - reproducing the items from the abstract on to bill paper in draft form ready for typing.

It may be that some offices still adopt this system of taking off and working-up as they are commonly called but they cannot be in the majority. In any case there is less need for the preparation of the abstract in civil engineering work as in building. For example, in a school all the painting dimensions for every room are added together on the abstract and stated in the bill as the total for the whole project. In a sewage treatment works, however, the work will usually be presented as a series of locational sub-bills each containing similar items; e.g. the Inlet Works, Primary Treatment Tanks will each be billed separately and will contain similar items. The adoption of this system greatly increases the efficiency of the post-contract administration.

An experienced civil engineering taker-off can usually take off in bill order and if he adopts a system of allocating only one item to each dimension sheet it removes the need for abstracting.

Conversely, some practices have adopted a system of writing full descriptions on the abstract sheets in bill order (a skill possessed by an experienced worker-up) and typing the bill direct from the abstract.

In the last 30 years most quantity surveying practices have adopted the cut-and-shuffle method. This comprises the writing of item descriptions and dimensions on to sensitised paper to produce two copies. When the taking off and squaring is complete the copies are split or 'cut' and one copy 'shuffled' into bill order with all sheets for the same item pinned together and their totals collected to produce a final quantity. More recently, other systems have come into use where the taker-off enters dimensions into a computer (sometimes by using a digitiser) which will then perform the squaring, abstracting, billing and printing functions.

Dimension paper

The ruling of dimension paper should conform to the requirements of BS3327 - Stationery for Quantity Surveying, and the paper is vertically separated into two parts by a double line each with four columns (Figure 1).

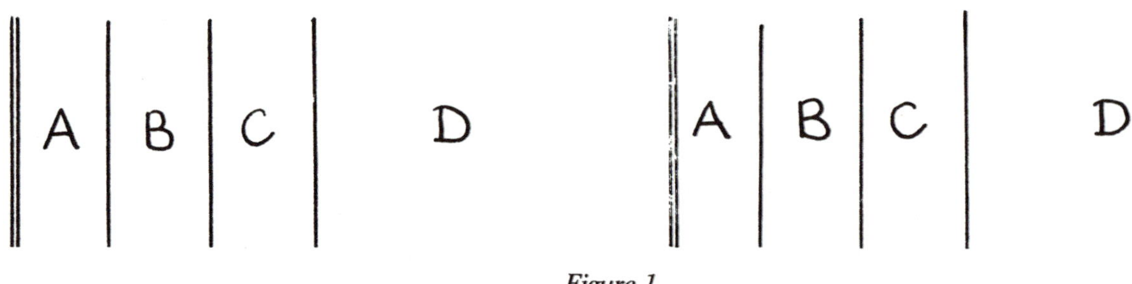

Figure 1

Column A is the 'timesing' and 'dotting on' column where multiplication and addition of the dimensions can be recorded (Figure 2).

A	B	C	D
5/	2	10	2 multiplied by 5
3.5/	2	16	2 multiplied by the sum of 3 and 5
5/	6.00 7.00	210.00	5 superficial areas with sides 6m and 7m long
2/3.5/	6.00 7.00	672.00	A superficial area with sides 6m and 7m long multiplied by the sum of 3 and 5 and further multiplied by 2

A	B	C	D
π/	2.00 2.00	12.57	a circle with a 2 metre radius multiplied by pi (3.142)
4/½/	4.00 2.00	16.00	Four triangles with base 4 metres and height 2 metres multiplied by ½ to produce the area.

Figure 2

The practice of 'dotting-on' should be used only where absolutely necessary because of the dangers of mistaking the dot for a decimal point.

Column B is the dimension column and receives the measurements taken off from the drawings. The dimensions are normally expressed to two decimal points (Figure 3).

A	B	C	D
	6	6	This represents an item which is repeated 6 times
6/	1	6	The same item can be expressed as 6 times 1
	8·00	8·00	Length of 8 metres
	8·00 2·00	10·00	Lengths of 8 and 2 metres added together
	8·00 2·00	16·00	Area of 16 square metres with sides of 8 and 2 metres

A	B	C	D
	8·00 2·00	16·00	Two areas totalling 31 square metres
	3·00 5·00	15·00	
		31·00	
	8·00 2·00 3·00	48·00	A volume of 48 cubic metres consisting of length 8 metres width 2 metres and depth of 3 metres
	8·00 2·00 3·00	48·00	Two volumes totalling 66 cubic metres.
	3·00 3·00 2·00	18·00	
		66·00	

Figure 3

It is important to note that it is the insertion of the horizontal line which determines whether the dimension is intended as a linear, superficial or cubic measurement (Figure 4).

A	B	C	D
	8.00		The lines separating the dimensions
	2.00		
	3.00		
		13.00	indicate three separate linear measurements totalling 13 linear metres.

A	B	C	D
	8.00		The absence of lines between
	2.00		
	3.00	48.00	the dimensions indicates a volume

Figure 4

The dimensions should always be recorded in the order of length, width and height. Column C is the squaring column where the result of the addition, subtraction or multiplication of the entries in the dimension column is recorded. Figures which are to be added or subtracted are bracketed together in the manner shown.

Deductions are sometimes necessary where it is easier to take an overall measurement and deduct the parts not required (Figure 5).

Column D is the description column where the item being measured is described. This is done by using a form of standard abbreviations which have been listed separately. This column also contains annotations giving the location of the dimensions and waste calculations which show the build up of the figures entered in the dimension column. (Figure 6).

Quite often two item descriptions share the same measurement and this is indicated by linking the descriptions with an ampersand.

It may be considered desirable to insert the appropriate CESMM 3 code in the description column as shown in Figure 6, but the value of doing this will depend upon the subsequent method of processing the dimensions and descriptions that is adopted.

A	B	C	D	A	B	C	D
			2m				
			3m				
			6m				
			10m				
	10.00		The deduction				
	6.00	60.00	dimensions are				
Ddt			recorded in the				
3.00			timesing column				
2.00		6.00	for convenience				
		54.00					
			or				
	10.00						
	6.00	60.00	The dimensions are recorded and taken to the abstract separately e.g.				
			Ddt ditto				
	3.00						
	2.00	6.00					

Figure 5

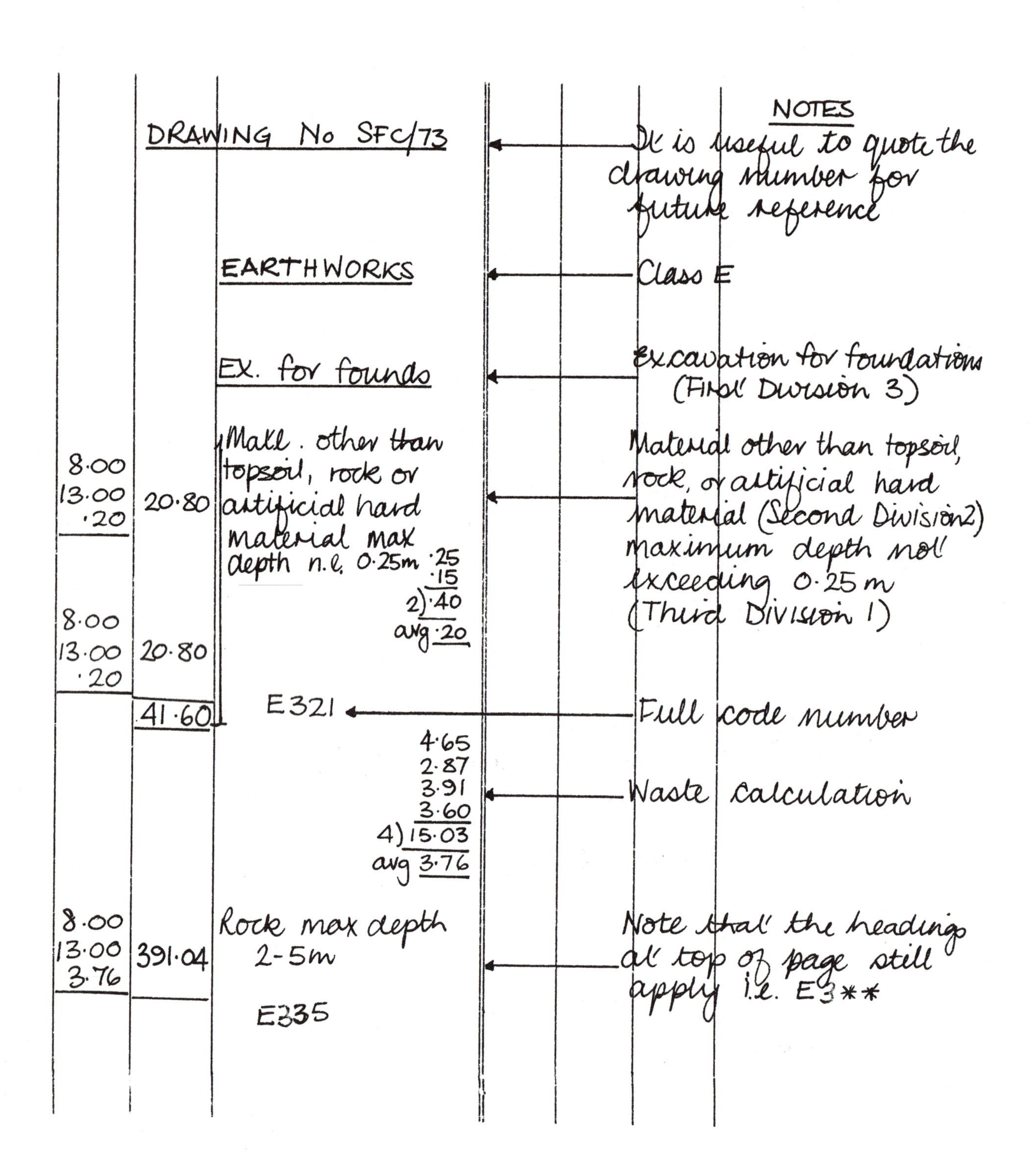
NOTES
DRAWING No SFC/73
It is useful to quote the drawing number for future reference
EARTHWORKS
Class E
EX. for founds
Excavation for foundations (First Division 3)
Matl. other than topsoil, rock or artificial hard material max depth n.e. 0·25m
8·00
13·00
·20
20·80
·25
·15
2)·40
avg ·20
Material other than topsoil, rock, or artificial hard material (Second Division 2) maximum depth not exceeding 0·25 m (Third Division 1)
8·00
13·00
·20
20·80
41·60
E321
Full code number
4·65
2·87
3·91
3·60
4)15·03
avg 3·76
Waste calculation
8·00
13·00
3·76
391·04
Rock max depth 2-5m
E335
Note that the headings at top of page still apply i.e. E3**

Figure 6

Dimensions	Squaring	Description	Note
		N.W Corner	Locational note
4·00 6·00 ·28	6·72	Conc. exposed at the Comm. Surf. Max. depth 0·25 – 0·5m	Vertical line drawn to link measurements to be added or deducted
1·00 1·00 ·28	0·28 7·00	E142	
9·50 3·50	33·25	Facing bkwk a.b.d. 215mm nom. thickness vert. st. walls; Eng. bond U221.1	
		&	Ampersand joining two descriptions means that they both share the same dimension
		Surface features; fair facing U278	
9·50		Ancillaries, dpc, bitumen, width 205mm U282·2	See paragraph 4.7

Figure 6 (continued)

Comm. brick BS 3921 size 215 × 102.5 × 65mm, jtg. in ct.m. (1:3), English bond, f.p. in ct.m. (1:3)

Sq.
215mm nom. thickness vert'. st. walls

1.	74.67		Dedt
2.	12.00	7.	14.20
	86.67	9.	2.40
Dedt	16.60		16.60
	70.07		
	= 70m²		

Sq.
215mm nom. thickness vert' face to concrete

3.	28.29
4.	18.20
	46.49
	= 46m²

Lin.
Surface features pilasters

			Dedt
7.	14.00		
7.	14.90	10.	12.00
8.	28.20		
9.	42.67		
	99.77		
Dedt	12.00		
	87.77		
	= 88m		

Ancillaries: damp proof courses; bitumen

Lin.
Width 102.5mm

8.	142.00
	= 142m

Ancillaries; built-in pipes and ducts

Nr.
Cross sectional area not exc. 0.05m²

14.	2
15.	3
	5
	= 5Nr

Nr.
Cross sectional area 0.08m²

14.	2
	= 2Nr

Figure 7

Abstracting

The skill of preparing an abstract lies in the ability of the worker-up to arrange the items abstracted from the dimension sheets in bill order. This may not seem too difficult a task to anyone who has not tried it, but when tender documents are being prepared in a rush against a tight deadline (which must be 99% of the time!) the worker-up may be handed the dimension sheets in small lots but must lay out his abstract to accommodate items he has not yet seen.

A typical abstract is set out in Figure 7. The figures on the left-hand side are the column numbers of the dimension sheets and the first item has been stroked through to indicate that it has been transferred to the draft bill.

Chapter 2
CESMM3 – HOW IT WORKS

SECTION 1: DEFINITIONS

Reference should be made to the Method of Measurement when considering the following notes:

1.1 All the words and expressions used in the Method of Measurement and in the Bills of Quantities are deemed to have the meaning that this section assigns to them.

1.2 Where reference is made to the Conditions of Contract, it means the ICE (6th Edition) Conditions of Contract issued in January 1991.

1.3 Where words and expressions from the Conditions of Contract are used in CESMM 3 they shall have the same meaning as they have in the Contract.

1.4 Where the word 'clause' is used it is referring to a clause in the Conditions of Contract. The word 'Paragraph' refers to the numbered paragraphs in Sections 1 to 7 inclusive of CESMM 3.

1.5 The word 'work' is defined in a broader sense than that in common usage to include not only the work to be carried out but also the labour materials and services to achieve that objective and to cover the liabilities, obligations and risks that are the Contractor's contracted responsibility as defined in the Contract.

1.6 The term 'expressly required' refers to a specific stated need for a course of action to be followed. This would normally take the form of a note on the drawings, a statement in the specification or an order by the Engineer in accordance with the appropriate clause in the Conditions of Contract.

1.7 The 'Bill of Quantities' is defined as a list of brief descriptions and estimated quantities. The quantities are defined as estimated because they are subject to admeasurement and are not expected to be totally accurate due to the unknown factors which occur in civil engineering work.

1.8 'Daywork' refers to the practice of carrying out and paying for work which it is difficult to measure and value by normal measurement conventions.

1.9 'Work Classification' is the list of the classes of work under which the work is to be measured in Section 8, e.g. Class E Earthworks.

1.10 'Original Surface' is defined as the ground before any work has been carried out. It should be stressed that this definition refers not to virgin untouched ground but ground on which no work has been carried out on the contract being measured.

1.11 'Final Surface' is the level as defined on the drawings where the excavation is completed. Pier or stanchion bases, soft spots or any other excavation below this level would be described as 'below the Final Surface'.

1.12 and 1.13 The definitions of 'Commencing Surface' and 'Excavated Surface' in CESMM 1 were amended significantly in CESMM 2.

These amendments were made because of the confusion surrounding their definitions. It is important when considering these changes that Rules M5, D4 and A4 of Class E are also considered.

Where only one type of material is encountered in an excavation the Commencing Surface is always the top surface prior to excavation and the Excavated Surface is always the bottom surface after excavation unless separate stages of excavation are expressly required. Complications arise, however, when the excavation penetrates through more than one type of material.

Previously, many takers-off were incorrectly stating the maximum depth of each *layer* of material instead of the maximum depth of the excavation itself.

The additional sentences added to Paragraphs 1.12 and 1.13 clarified what was always intended, i.e. that the top of the excavation is the Commencing Surface and the bottom is the Excavated Surface no matter how many layers of different materials lie between the two.

The maximum depth of the excavation of an individual layer of material in accordance with Paragraph 5.21 is the maximum depth of the excavation even though the depth of the layer is significantly less. This rule is acceptable when the excavation is a regular shaped hole in the ground but problems arise when it is applied to a large area of land-forming or reduced level excavation.

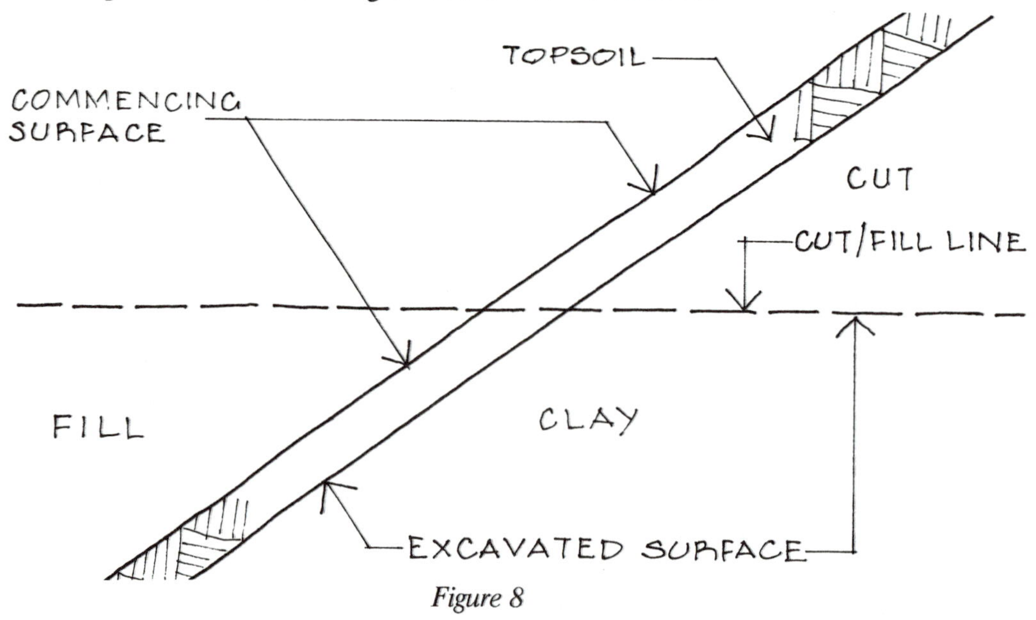

Figure 8

Figure 8 illustrates a sloping site which is being cut and filled to produce a level surface with the topsoil removed separately. The first operation to remove the topsoil is where the anomaly occurs. The information which is of most interest to the estimator is the depth of the topsoil itself. Over the filled area the Commencing Surface and the Excavated Surface is the top and bottom of the topsoil layer and the maximum depth in accordance with Paragraph 5.21 is the difference between the two.

Unfortunately this is not the case over the cut area. The Excavated Surface is the cut-and-fill line so the maximum depth is far greater in this area than over the filled section. This creates two separate items for identical work. This is obviously undesirable because the estimator requires only one item to cover all the work to be carried out under the scope of the description. Careful thought should be given to including either an additional preamble or enlarging the description to cover this set of circumstances.

1.14 A 'hyphen between two dimensions' means the range of dimensions between the figures quoted in excess of the first number but including the second, e.g. '150 - 300mm' means 151 to 300mm inclusive.

Other words commonly used in CESMM 3 are defined as follows:

'shall' - mandatory
'should' - optional
'may' - optional
'given' - stated in the Tender Documents
'inserted' - included by the Contractor

1.15 Problems arose in the last few years over the reference to BS numbers where the documents were intended for use in countries belonging the European Community, particularly Ireland. This has been overcome by the widening of the clauses' meaning to include equivalent standards of other countries.

SECTION 2: GENERAL PRINCIPLES

2.1 The formal title of the document is 'Civil Engineering Standard Method of Measurement' which can be referred to in an abbreviated form as CESMM. Although not mentioned it is inevitable that the revised edition will be called CESMM 3 and the previous versions CESMM 1 and 2.

2.2 Building (with the exception of work covered by Class Z), mechanical, electrical or any other work which is not civil engineering, but which is part of a civil contract should be measured in accordance with the appropriate method of measurement for the particular work involved. An item should be included in the Preamble to the Bill of Quantities stating which work is affected and how it has been measured.

2.3 Although this paragraph states that the defined procedures of the CESMM 3 shall be observed in the preparation and pricing of the Bill of Quantities and the description and measurement of the quantities and items of work, it is possible to depart from the rules where thought appropriate. The authority for the departure lies in Paragraphs 5.4 and 5.10.

2.4 The object of preparing the Bill of Quantities is stated as twofold. First, to assist estimators to produce an accurate tender efficiently. It should be borne in mind, however, that the quality of the drawings plays a major part in achieving this aim by enabling the taker-off to produce an accurate bill and also by allowing the estimator to make sound engineering judgements on methods of working. Second, the Bill of Quantites should be prepared in such a style (within the framework of the CESMM 3) to assist the post-contract administration to be carried out in an efficient and cost-effective manner.

2.5 This paragraph defines the need to present the measured items in the Bill of Quantities in sufficient detail so that items covering separate classes of work can be easily distinguished. It also requires that work of the same nature carried out in different locations is kept separate. This is a direct result of complying with the requirements of the preceding Paragraph 2.4 and is intended to assist the site surveyor or measurement engineer in the admeasurement and valuing of the work.
For example if the Bill of Quantities is being prepared for a water treatment works, it is desirable that separate parts are given for, say, the raw water storage reservoir, the slow sand filters and pump houses. The alternative of adding together similar items from each structure would produce a Bill of Quantities which would have limited post-contract value.

2.6 This paragraph states that all work (as defined in Paragraph 1.5) that is expressly required (as defined in Paragraph 1.6) should be covered in the Bill of Quantities. This paragraph is intended to remove any doubt about the status of formwork or any other temporary works which are required but are not left on site on completion to become the property of the Employer.

2.7 It is the proper application of the Work Classification tables that enables the aims of Paragraphs 2.4, 2.5 and 2.6 to be achieved. The tables and rules state how the work is to be divided, the scope of the item descriptions, the measurement unit for each individual item and the method (with a small 'm') to be adopted to produce the quantities.

SECTION 3: APPLICATION OF THE WORK CLASSIFICATION

This section deals with the details of how to use the Work Classification and the Rules. There are 26 classes in the Work Classification and each class comprises:

(a) up to three divisions each containing up to a maximum of eight descriptive features of the work

(b) units of measurement

(c) Measurement Rules

(d) Definition Rules

(e) Coverage Rules

(f) Additional Description Rules.

3.1 The Work Classification is prefaced by an 'Includes' and 'Excludes' section which defines in general terms the nature and scope of the work contained in each individual class This is particularly useful where there would appear to be a choice of which class to use. For example, in Class R (Roads and Pavings) it clearly states that associated Earthworks and Drainage are to be measured separately in Class E and Classes I, J, K and L respectively.

There are three divisions or levels of description and up to eight part descriptions or descriptive features in each division. These descriptive features are intended to cover the broad range of activities in civil engineering but they are not exhaustive. At its simplest, it should be possible to take a descriptive feature from each division and produce an item description, e.g. in Class C - Concrete Ancillaries the following item description could be assembled:

First Division 2 Formwork: fair finish

Second Division 3 Plane battered

Third Division 1 Width: not exceeding 0.1m.

This item would then appear in the bill as 'Formwork fair finish, plane battered, width not exceeding 0.1m' with a coding of C. 2 3 1.

This example is straightforward and it is not typical of the compilation of most item descriptions. It is essential that the rules are carefully studied before attempting to build up a description in case there is a restriction which would not be apparent by merely assembling descriptive features from the divisions. The use of the exact form of wording in the divisions is not mandatory but it would be unwise to depart from the printed descriptive features without a sound reason.

It is not essential that the punctuation is adhered to rigidly. The Work Classification is intended as a foundation upon which the 'taker-off' can base the needs of the individual project he is working on.

In the selection process from the three divisions it is important to note the need to observe the function of the horizontal lines. The selection must be made horizontally and be contained within the lines joining adjacent divisions. It would not be possible, for example, to select features in Class G to produce an item coded G. 1 1 8 because of the line preventing the Second Division Code 1 being linked with any of the void depth descriptive features.

3.2 This paragraph states that there is a basic assumption built into the descriptive features which removes the need for a comprehensive list of activities which the Contractor must perform in order to achieve the fixing of the material described. For example:

Item U. 1 2 1 does not require a preface stating that the rate set against the item must include for unloading the bricks, transporting them to a suitable place, carrying them to the Bricklayers' side, lifting them by the Bricklayers' hand and laying to the correct line and level. That is all assumed in the item 'Common brickwork, thickness 230mm, vertical straight wall'.

3.3 If the scope of the work to be carried out is less than that normally covered by item descriptions in similar circumstances then that limitation must be clearly defined. For example, it is not uncommon for the Employer to purchase special pipes or fittings in advance of the main civil contract where there is a long delay between order and delivery.

These materials would be handed to the Contractor on a 'free issue' basis and the item description would be headed 'Fix only'.

The scope of the term 'Fix only' should be clearly defined in the Preamble and could include such activities as taking charge of, handling, storing, transporting, multiple handling, laying and jointing including all necessary cutting.

The item or items involved must be unequivocally stated to avoid any ambiguities and if a number of items are covered by the limitation it would be prudent to insert 'End of fix only' after the last item.

3.4 The taker-off must not take more than one descriptive feature from any one division when compiling an item description.

3.5 The units of measurement are stated within the Work Classification and apply to all the items to which the descriptive features relate.

3.6 The Measurement Rules are defined as the circumstances existing for the implementation of the rules listed under M1 *et seq*. Reference must be made to Paragraph 5.18 for general additional information on measurement conventions.

3.7 Definition Rules lay down the parameters of the class of work covered by words or phrases in the Work Classification or the Bill of Quantities.

3.8 Coverage Rules describe the scope of work that is included in an item description although part of the required action may not be specifically mentioned. The rule will not necessarily cover all the work required and does not cover any work included under the Method Related Charges section.

3.9 The Additional Description Rules are included to provide a facility for the inclusion of extra descriptive features where those listed in the Work Classification are not considered comprehensive enough. The authority for this comes from Paragraph 3.1.

3.10 This note clears any confusion between the Work Classification and the Additional Description Rules. The example quoted refers to Class I. If item I. 5 2 3 was assembled by reprinting the descriptive features, the items would read:

'Clay pipes, nominal bore 200 - 300mm, in trenches depth 1.5 - 2m.'

Rule A2 requires, among other things, that the nominal bore is stated. Paragraph 3.10 states that where these circumstances occur the provision of the note should override that of the descriptive feature so this item would read:

'Clay pipes, nominal bore 250mm, in trenches depth 1.5 - 2m.'

N.B. The other requirements of A2 would probably be included in a heading or in an enlargement of the First Division entry 'Clay pipes'.

3.11 It should be noted that on the rules side of the page in many classes there is a horizontal double line near the top of the page. All the rules above this line refer to all the work in the class. Rules below the double line refer only to the work contained within the same horizontal ruling.

SECTION 4: CODING AND NUMBERING OF ITEMS

This section deals with coding of items and it is important to remember that the provisions of the section are not mandatory. The value of applying the CESMM system of coding must be judged by the engineer or surveyor preparing the Bill of Quantities. If by following the recommendations of this section a series of unwieldy codings is produced it may be better not to apply them.

The aim of the coding is to produce a uniformity of presentation to assist the needs of the estimator and the post-contract administration.

4.1 The structure of the coding system is simple and easy to apply. Each item is allocated four basic symbols to produce a four-unit code. The class is represented by the class letter (e.g. G - Concrete Ancillaries) followed by the numbers taken from the First, Second and Third Division, respectively.

Item G.1 2 3 therefore refers to an item description of 'Formwork, rough finish plane sloping, width 0.2 - 0.4m'.

4.2 Where the symbol * appears it denotes all the numbers in the appropriate division. For example item G. 1 * 3 refers to items G. 1 1 3, G. 1 2 3, G. 1 3 3, G. 1 4 3 and G. 1 5 3. This symbol would never appear in the coding of an item description in a Bill of Quantities because by definition it refers to more than one item. Its main use is to assist, in an abbreviated form, in the application of the rules (e.g. see Rule M2 in Class P).

4.3 The option is given whether to apply the provisions of this section in the Bill of Quantities or not. The authors' reservations stated at the beginning of this section about the use of the codes apply only to their appearance in the Bill of Quantities. It is desirable that they are used in the taking off and working-up stages as an aid to the presentation of the items in a regular and uniform bill order.

4.4 This paragraph states that if the code numbers are to appear in the Bill of Quantities they must be placed in the item number column and not become part of the item description. The code numbers have no contractual significance.

4.5 The highest number listed in the Work Classification is 8 and if a completely new descriptive feature is to be added it should be given the digit 9 in the appropriate division.

4.6 Conversely, the digit 0 should be used if no descriptive feature in the Work Classification applies or if there are no entries in the division itself.

4.7 The code numbers refer only to the descriptive features in the three divisions. If additional descriptive features are required (see Paragraph 3.9) by the implementation of the Additional Description Rules they shall be identified by the addition of a further digit at the end of the code number. The example quoted in Paragraph 4.7 quotes item H. 1 3 6, but when Rule A1 is applied the additional information generates a code number of H. 1 3 6.1. If there was a need for more than one item they would appear as H. 1 3 6.2, H. 1 3 6.3, H. 1 3 6.4, etc.
What is not explained is the technique for dealing with this situation when the number of items containing additional descriptive features exceeds 9. The choice lies between H. 1 3 6.10, H. 1 3 6.9.1 or H. 01 03 06.01, but the authors feel that the latter nomenclature is probably the most suitable.

SECTION 5: PREPARATION OF THE BILL OF QUANTITIES

It should be noted that the term used in this section heading, i.e. 'Bill of Quantities', is correct. The phrase 'Bills of Quantities' is more appropriate to a building contract where the General Summary contains a list of individual Bills. In civil engineering documents the equivalent Bills are called Parts (Paragraph 5.23) so the overall document is a Bill of Quantities.

5.1 This paragraph states that the rules and provisions used in the pre-contract exercise of measuring the work also apply to the post-contract task of measurement. The correct term for this task is re-measurement where the work is physically measured on site or admeasurement where the actual quantities are calculated from records.

5.2 There are five sections in the Bill of Quantities:

A List of Principal Quantities

B Preamble

C Daywork Schedule

D Work Items (divided into parts)

E Grand Summary

The Daywork Schedule can be omitted from the Bill of Quantities if required. The above sections should be allocated the letters A to E and the parts of the Bill contained within Section D are enumerated, e.g.

Section A List of Principal Quantities

B Preamble

C Daywork Schedule

D Work Items
- Part 1 General Items
- Part 2 Boldon Sewers
- Part 3 Cleadon Sewers
- Part 4 Rising Mains
- Part 5 Shackleton Pumping Station
- Part 6 Roker Pumping Station

E Grand Summary

5.3 It should be noted that the list of principal quantities is prepared by the taker-off or the person assembling the Bill of Quantities and should satisfy two requirements. First, to give the estimator an early feel for the scope of the work before he commences pricing and, second, to assist the participants at the Contractor's pre-tender meeting with regard to the type and size of the contract when considering the application of the adjustment item (Paragraph 5.26). The list has no contractual significance. A notional list of principal quantities for the job mentioned in Paragraph 5.2 could be as follows:

Part	Item	Quantity
Part 1 General Items		
	Provisional Sum	75,000
	Prime Cost Sums	100,000
Part 2 Boldon Sewers		
	Pipelines	1200m
	Manholes	22nr
Part 3 Cleadon Sewers		
	Pipelines	1500m
	Manholes	28nr
Part 4 Rising Mains		
	Pipelines	2000m
	Valve Chambers	12nr
Part 5 Shackleton Pumping Station		
	Excavation	600m3
	Concrete	50m3
	Brickwork	200m2
Part 6 Roker Pumping Station		
	Excavation	650m3
	Concrete	80m3
	Brickwork	240m3

5.4 The Preamble is an extremely important section of the Bill of Quantities and is the potentially vital source of information to the estimator. If any other Methods of Measurement have been used in the preparation of the Bill of Quantities, the fact should be recorded here. This is not uncommon where, say, the Administration Building of a Sewage Treatment Works or even the superstructure of a large pumping station has been measured in accordance with the current Method of Measurement for Building Works, although the inclusion of Class Z should reduce the need for this.

The Preamble should also contain information on work to be designed by the Contractor or where the Contractor is involved in alternative forms of construction. The style of measurement to deal with these events will involve a departure from the rules laid down so warranting an insertion in the Preamble.

The Preamble will also contain a list of departures from the rules and conventions of CESMM 3 if the taker-off considers it desirable. Because the preamble note will usually commence 'Notwithstanding the provisions of...' these notes have become known as 'notwithstanding' clauses. A common example affects Paragraph 5.9. Many surveyors and engineers do not wish to adopt the lining-out system as set out in this paragraph and would insert the following clause in the Preamble:

'Notwithstanding the provisions of Paragraph 5.9, lines have not been drawn across each bill page to separate headings and sub-headings.'

It should be noted that the Preamble can also be used as a vehicle to extend the Rules. The Item Coverage Rules are the most likely to be enlarged and the taker-off should not hesitate to use this facility in order to improve the quality of the information provided to the estimator.

Where the word 'Preamble' is used in this book it refers to the section of the Bill of Quantities as defined in this paragraph. The term 'preamble' (with a small p) has been used to mean a clause or note.

5.5 It is also necessary to include in the Preamble a definition of rock. On first consideration it may seem odd that what is primarily an engineering matter should find its place in the Preamble of the Bill of Quantities. The reason of course is that it is the definition of what rock is, that will determine to what extent it is measured. It is the measurement of rock which is the main consideration in the Preamble and the definition should clearly state in geological terms what materials will be defined and paid for as rock. If any borehole information is available it would be useful to make reference to the logs and use the same terms wherever possible.

The practice of defining rock as 'material which in the opinion of the Engineer can only be removed by blasting, pneumatic tools, or wedges' is not recommended because it creates doubt in the minds of the taker-off, the estimator and, most importantly, of the people engaged the post-contract work.

5.6 It is not mandatory that a Daywork Schedule is included in the Bill of Quantities. If it was omitted, either by design or error, any daywork that occurred would be measured in accordance with Clause 52(3) of the Conditions of Contract and valued at the rates applicable to the FCEC schedules without any increase or decrease to the current percentages.

The other two methods of including dayworks in the Bill of Quantities are fully described in Chapter 3 - General Items.

5.7 Where the method set out in Paragraph 5.6(6) is adopted for dayworks it is usual to include separate provisional sums for the Labour, Materials, Plant and Supplementary Charges. The Contractor would be given the opportunity to insert his adjustment percentages after each item. (Chapter 4 - General Items).

5.8 It is important that careful thought is given to layout of the Bill of Quantities. Almost the first task of the taker-off should be to consult the Engineer and draw up the Grand Summary to identify the various parts. In the example given in Paragraph 5.2 the various parts are easily identified. The work in Part 2 headed Boldon Sewers should be presented in a style which locates the work in more detail, e.g. Manhole 1 to Manhole 2, etc.

In sewage disposal works and water treatment works it is usually quite straightforward to prepare a list of parts based on individual structures in the same order in which they are involved in the treatment process. It is more difficult in major bridge contracts and it is usual for the parts to be related more to CESMM 3 work classes than the locations of the work.

Whatever decisions are taken regarding the arrangements of the parts, the order of billing within each part should conform to the order of classes and items created by the Classifications within each class.

5.9 This paragraph provides for the placing of headings and sub-headings above item descriptions to prevent the repetition of material common to each item. These headings and sub-headings should be repeated at the top of each new page (perhaps in an abbreviated form) to assist the estimator in appreciating the full content of the item he is pricing.

A more controversial part of this paragraph deals with the procedure of what has come to be known as 'lining out'. This is the arrangement by which lines are drawn across the Item Description column of the bill page to end the influence of a previous heading or sub-heading.

Figure 9 shows how the lining out is done.

5.10 This paragraph gives the authority to the taker-off to add additional descriptive material to a description constructed from the three divisions if the work being measured has special characteristics which 'give rise to special methods of construction or consideration of cost'.

The implications of this paragraph are far reaching for the taker-off. He can impose his own judgement on the measurement of any item and depart from the format provided he believes it is a special case. It would be unwise, however, for the taker-off to abuse the power entrusted to him by this paragraph. The larger the number of items in the Bill of Quantities that conform to the preferred style of the CESMM 3 the more uniformity will be achieved which will benefit all parties. The taker-off should use the powers of this paragraph sparingly but on the occasions where it is felt that a new form of item description or additional descriptive material is necessary, the opportunity should be taken with the needs of the estimator and post-contract administration overriding those of the generalities of CESMM 3.

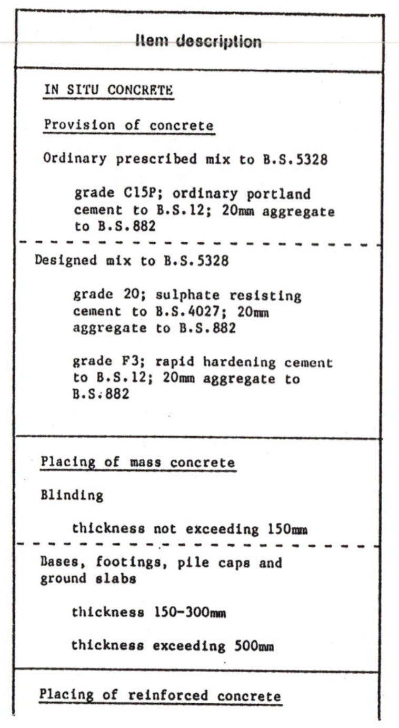

Item description
IN SITU CONCRETE Provision of concrete Ordinary prescribed mix to B.S.5328 grade C15P; ordinary portland cement to B.S.12; 20mm aggregate to B.S.882 - Designed mix to B.S.5328 grade 20; sulphate resisting cement to B.S.4027; 20mm aggregate to B.S.882 grade F3; rapid hardening cement to B.S.12; 20mm aggregate to B.S.882
Placing of mass concrete Blinding thickness not exceeding 150mm - Bases, footings, pile caps and ground slabs thickness 150-300mm thickness exceeding 500mm
Placing of reinforced concrete

Strictly speaking the dotted lines should form part of the lining out, because they are below the last item to which the heading and sub heading applies. However, to do so would mean that the underlined main headings would have to be repeated below the dotted lines and this is undesirable. In this instance the lining out only occurs below the last item to which the main heading applies.

Figure 9

If the new form of item description conflicts with the rules of the method, a 'notwithstanding' clause should be raised in the Preamble (Paragraph 5.4).

5.11 This paragraph reinforces the secondary role of the Bill of Quantities. The estimator is actively discouraged from relying on the item descriptions as a sole source for the information he requires to build up his rates. The 'exact nature and extent of the work' (or as near as it is possible to define it) must be determined from the Drawings, the Specification and the Contract. The item descriptions should not be held to be comprehensive but used to 'identify' the work being measured. This downgrading of the descriptions does not in any way relieve the taker-off from his responsibility of producing the most lucid descriptions he can within the framework of the method of measurement.

5.12 Where an unusual feature occurs in the work it is sometimes easier and more accurate to direct the estimator to a clause in the Specification or a detail of a drawing rather than produce a clumsy description which does not fully cover the work to be measured. Although this paragraph gives the authority for this form of referencing, it also contains an important proviso that the reference must be precise. A general reference to a drawing containing standard details would be unacceptable unless it identified the exact detail in the drawing being referred to.

5.13 In civil engineering contracts the work is subject to admeasurement. The quantities that are measured in the contract document are approximate because of the uncertainties inherent in civil engineering. The method of measurement has by necessity great flexibility and affords the taker-off opportunities to use his professional skill and judgement denied to his opposite number in the building side of the industry. This paragraph demonstrates this freedom. There are many situations where the choice of the style of measurement and placing of the items in the bill is entirely at the discretion of the taker-off. One example of this concerns thrust blocks in Class L. Where a block is large, say 10m3, the taker-off may feel it more helpful to measure it in detail using Classes E, F and G rather than item L. 7 8 0. A major consideration in this decision would be the knowledge that if the item was enumerated and the drawing reference given (as Paragraph 5.12) each tendering estimator would need to measure the excavation, concrete and formwork so it is less wasteful if the taker-off prepares these. Whatever decision the taker-off makes in these matters it should be clear and unequivocal so that the estimators are not confronted by ambiguities and uncertainties.

The comments made earlier about the quantities being approximate should not give the impression that anything less than the highest professional standards should be employed in the preparation of the tender documents. The quantities are described as approximate because in many cases the scope of the work is not known but the measurements should be as accurate as possible even in the knowledge that they will be taken again on completion of the works.

5.14 Where a range of dimensions is given in the Work Classification tables but the measured items have an identical thickness it is permissible to state the thickness instead of the range. For example, item K. 1 1 3 describes brick manholes in a depth range of 2 to 2.5m. If there were three manholes all 2.2m deep the item description should read 'Manholes, brick, depth 2.2m' and would carry the same code number K. 1 1 3.

5.15 Where work is to be carried out by a Nominated Sub-Contractor the estimated cost of the work should be given as a Prime Cost Item. Items to cover what used to be called general and special attendances follow this sum and are dealt with in Chapter 3 General Items.

The scope of the facilities to be available to the Nominated Sub-Contractor include for temporary roads, hoists and disposing of rubbish.

5.16 Any goods, materials or services supplied by a Nominated Sub-Contractor which are to be used by the Contractor must be referenced to the Prime Cost Item involved by a heading or mention made in the item itself.

5.17 The use of provisional quantities is discouraged by CESMM. Prior to 1976, items frequently appeared in Bills of Quantities under a heading of Provisional. This procedure was usually adopted because the Design Engineer either did not know the scope of the work or did not have enough time to design it. The assumption that the Contractor had better knowledge at tender stage than the Engineer, and was able to price the work, was completely unacceptable. On occasions, if the provisional quantities included were small, the Contractor would insert high rates which would hardly affect his tender total but could lead to a windfall if the quantities increased on admeasurement.

This paragraph states how the cost of uncertainties in design should be treated. If there are specific areas of work where the design has not advanced far enough to allow accurate quantities to be prepared, the work should be placed in the General Itemsagainst a Provisional Sum. It is also usual to include a Provisional Sum in the

Grand Summary for general contingencies.

Recently, however, some Employers are resisting the inclusion of this general contingency allowance in the spurious belief that the Contractor will somehow regard that sum as 'spendable' and attempt to recover it through claims. This notion shows little confidence in the skills of post-contract management team acting on the Employer's behalf.

5.18 This paragraph confirms the long-standing convention that measurements are taken net - unless there is a specific requirement to the contrary. Ideally, the quantities are computed from dimensions on the drawings. Common sense must be applied in the matter of rounding-off quantities. The total quantity and the effect on it of rounding off must be considered.

5.19 The units of measurement are set out in this paragraph and the abbreviations must be used in the Bill of Quantities. Care should be taken when using the abbreviation for Number because the handwritten 'nr' is very similar to 'm' and mistakes can be made when documents are produced at speed by confusing the two abbreviations.

5.20 It is a requirement that where a body of open water is either on the site or bounds the site, it shall be identified in the Preamble to the Bill of Quantities stating its boundaries and levels or fluctuating levels. This requirement should not be taken too literally. If a power station was to be constructed on the west Cornish coast it would be sufficient to state in the Preamble that the Atlantic Ocean was adjacent to the site together with tidal information. It would be unnecessary and foolish to attempt to define the bounds of the Atlantic!

It is interesting to note that Rule A2 in Class E provides a further requirement for the body of water to be identified in the item description for work which is below the feature; this requirement is not thought necessary in other Classes such as F, I or P where similar situations could occur.

5.21 This paragraph deals with the definition of the terms Commencing Surface and Excavated Surface. This matter has been dealt with under paragraphs 1.12 and 1.13. See also Class E.

5.22 A sample of the ruling and headings of bill paper is shown in Figure 10.

Number	Item description	Unit	Quantity	Rate	Amount	

Figure 10

5.23 The summary of each Part would be printed on standard bill paper but the Part total would be styled 'Carried to Grand Summary' (see Figure 11).

PART 4 - RISING MAINS

Number	Item description	Unit	Quantity	Rate	Amount	
	COLLECTION					
	Page 6/1					
	Page 6/2					
	Page 6/3					
	Page 6/4					
	Page 6/5					
	Page 6/6					
	Total Carried to Grand Summary £					

Figure 11

5.24 The Grand Summary collects the totals from the parts of the Bill of Quantities and is usually printed on plain paper (Figure 12).

GRAND SUMMARY

		£ p
PART 1 GENERAL ITEMS		
PART 2 BOLDON SEWERS		
PART 3 CLEADON SEWERS		
PART 4 RISING MAINS		
PART 5 SHACKLETON PUMPING STATION		
PART 6 ROKER PUMPING STATION		
		________
	£	
GENERAL CONTINGENCY ALLOWANCE		50,000 00
		________
ADJUSTMENT ITEM ADD/DEDUCT*		
		________
TENDER TOTAL	£	
		________

* Delete as required.

Figure 12

5.25 The General Contingency Allowance is discussed in Paragraph 5.17.

5.26 The Adjustment Item is to be placed at the end of the Grand Summary and its significance and purpose are discussed in Paragraph 6.3, 6.4 and 6.5.

5.27 The Grand Summary must contain a provision for the addition of the individual bill parts, the General Contingency Allowance and the addition or subtraction of the Adjustment Item. This total is often called the Tender Total but it should not strictly receive that title until the acceptance of the Contractor's Tender for the Works in accordance with Clause 1(i)(h).

SECTION 6: COMPLETION, PRICING AND USE OF THE BILL OF QUANTITIES

6.1 The rates to be inserted in the rates column shall be expressed in pounds sterling with the pence given as a decimal fraction. Thus 6.47 denotes 6 pounds 47 pence. It is important that the amount is written clearly with the decimal point well defined to avoid subsequent misunderstandings and disputes. If 647 was entered in the rate column and it was intended to be 647 pounds it should be expressed as 647.00. Careful inspection of the presentation of the rates together with their values should be part of the tender appraisal process. Where rates are not inserted the other priced items are deemed to carry the price of the unpriced items.

6.2 This paragraph confirms the requirement made in Paragraph 5.22 that each part must be totalled and then carried to the Grand Summary.

6.3 and 6.4 The introduction of the Adjustment Item was warmly welcomed by the industry in 1976 when CESMM 1 was published and its use is now well established. Most Contractors contend that they are rarely allowed sufficient time to prepare their tenders. Each job needs careful scrutiny and the application of sound engineering judgements to determine how the construction work should be tackled. Enquiries for material prices and sub-contractors' quotations' must be sent out and it frequently happens that they do not arrive until quite late in the tender period.
If, for example, a quote for ready-mixed concrete was obtained on the day before a tender was due to be submitted which was substantially below other quotations the Contractor would be keen to include the effect of the offer in his tender.

Pre-1976 he would have probably deducted the difference from a convenient sum in the General Items and thus created an imbalance in the pricing structure. By using the Adjustment Item the Contractor can now increase or decrease his Tender Total at a stroke yet still present a well-balanced bid.

Another reason for using the device could arise from the Contractor winning or losing other contracts during the tender period which would lessen or increase his determination to put in a keen bid. This decision would normally be taken at a tender appraisal meeting before the signing of the offer.

The sum inserted should be regarded as a lump sum and will be paid or deducted in instalments in the same proportion that the amount being certified bears to the Tender Total before the application of the Adjustment Item in the Grand Summary.

It is a requirement of CESMM that this should be stated in the Preamble to the Bill of Quantities.

The amount involved shall be calculated before the deduction of retentions and the aggregate total must not exceed the amount inserted in the Grand Summary. When the Certificate of Substantial Completion (Clause 48) is issued the difference (if any) between the aggregate total and the amount in the Grand Summary should be paid or deducted in the next certificate to be issued.

6.5 This new paragraph clears up any misunderstandings over the position of applying the Adjustment Item when the Contract is subject to a Contracts Price Fluctuation (CPF) clause. When the Effective Value is calculated it should take into account the effect of deducting or adding the Adjustment Item as appropriate in assessing the amount due to the Contractor under Clause 60.

SECTION 7: METHOD-RELATED CHARGES

Method-Related Charges were first introduced in CESMM 1 in 1976. It was felt that a different approach was required in the valuation of items where quantities were increased or decreased from those in the tender document. Research had shown that modern construction techniques had substantially increased the proportion of the non-quantity related part of a Contractor's costs to a level where it was becoming inequitable both to the Employer and the Contractor that changes in the quantities should be valued merely by multiplying the admeasured quantity by the bill rate.

The unit rates are made up of quantity-related costs - the labour, material and that part of the plant and overheads directly related to the item of work being constructed, and the non-quantity related items such as the transporting to site, erection, maintenance, dismantling of plant, cabins and other consumables which may have no direct link with the quantity of the permanent works being constructed.

It is sensible therefore to give the Contractor the opportunity to declare the cost of those items which he does not wish to be subject to the admeasurement process so that his real costs are recovered without being affected by changes in quantity.

7.1 A Method-Related Charge is the sum inserted in a Bill of Quantities in the space provided (Class A) and is either a Time-Related Charge or a Fixed Charge.

A Time-Related Charge is a sum which is directly proportional to the time taken to carry out the work which is described.

A Fixed Charge is a sum which is neither quantity-related nor time-related but is a set cost regardless of changes in the admeasured work or the time taken to execute it, e.g. the cost of bringing a batching plant on to site.

7.2 The Contractor has the opportunity to insert the cost of Time-Related and Fixed Charges in the Bill of Quantities (see General Items Class A).

7.3 The Contractor should enter the item description for his Method-Related Charges in the same order as the order of classification in Class A. He must also list the Time-Related Charges separately from the Fixed Charges and insert a sum against each item. He has the freedom, of course, to enter other items which are not listed or do not have a direct counterpart in Class A.

7.4 The Contractor should unambiguously describe the scope of the work that is covered by each sum. He should also list the labour, plant and materials involved and, where applicable, state the parts of the Permanent or Temporary Works that are linked to the sum inserted.

7.5 The Contractor is not obliged to follow the method he has set out in the tender document when he carries out the work on site.

7.6 This paragraph states that the Method-Related Charges are not to be admeasured. The wording was expanded in CESMM 2 to include the words '. . . but shall be deemed to be prices for the purposes of Clauses 52(1), 52(2) and 56(2)'.

The addition of these words confirms what was always inferred in CESMM 1. It is sometimes difficult for students to understand the true meaning of what this paragraph covers. An unequivocal statement that Method-Related Charges are not to be admeasured seems to sit uneasily beside the assertion that they are subject to the provisions of Clause 56(2).

If the items for Time-Related Charges and Fixed Charges have been set out by the Contractor in a sensible fashion it should be a straightforward task of apportionment each month to arrive at the amount due. One complication may arise if the time being expended on a Time-Related Charge looks like increasing or decreasing from that shown in the Bill of Quantities. If, for example, an operation was scheduled to occupy 6 months, after 1 month the Contractor would rightly ask for 1/6 of the sum. If his progress increased dramatically and the work looked like being completed in only 4 months, he would be fully entitled to ask for 1/2 of the sum at the end of the second month. It can be seen, therefore, that the numerator in the fraction will increase each month by 1 but the denominator could vary as the Contractor and the Engineer determine the likely length of time the event will last.

It should be noted that a statement must be included in the Preamble to the Bill of Quantities confirming that payment must be made in accordance with Clauses 60(1)(d) and 60(2)(a). This apparent confliction is explained if one remembers that the charges will be paid in full whether they were incurred tenfold or not at all, providing the risk that the Contractor undertook and priced did not vary.

If there was a significant change in quantity or in the time an item of plant was required which was substantially different from that envisaged when the Contractor prepared his tender, then an adjustment to the Method-Related Charges would be in order and the provisions of Clauses 52(1), 52(2) and 56(2) would be implemented.

7.7 Method-Related Charges are to be certified and paid for in exactly the same manner as other parts of the work and this should be stated in the Preamble.

7.8 It may be that the method of working stated by the Contractor is not adopted (Paragraph 7.5) but in the absence of a variation (see Paragraph 7.6) the sum inserted must be paid in full. It is obviously desirable that the Contractor and the Engineer agree a method of apportioning the sum each month for payment by linking it to progress of a relevant part of the works or indeed the whole works. If agreement cannot be reached the sum would then be added to the Adjustment Item (which would increase a positive Adjustment Item and decrease a negative one) and would be treated as described in Paragraph 6.4.

SECTION 8: WORK CLASSIFICATION

This section lists the 26 classes in CESMM 3. Each class consists of the Classification Tables containing three divisions of descriptive features and four types of rules. See Section 3 for details on the application of the tables and rules.

Chapter 3
PUMPING STATION NO. 1

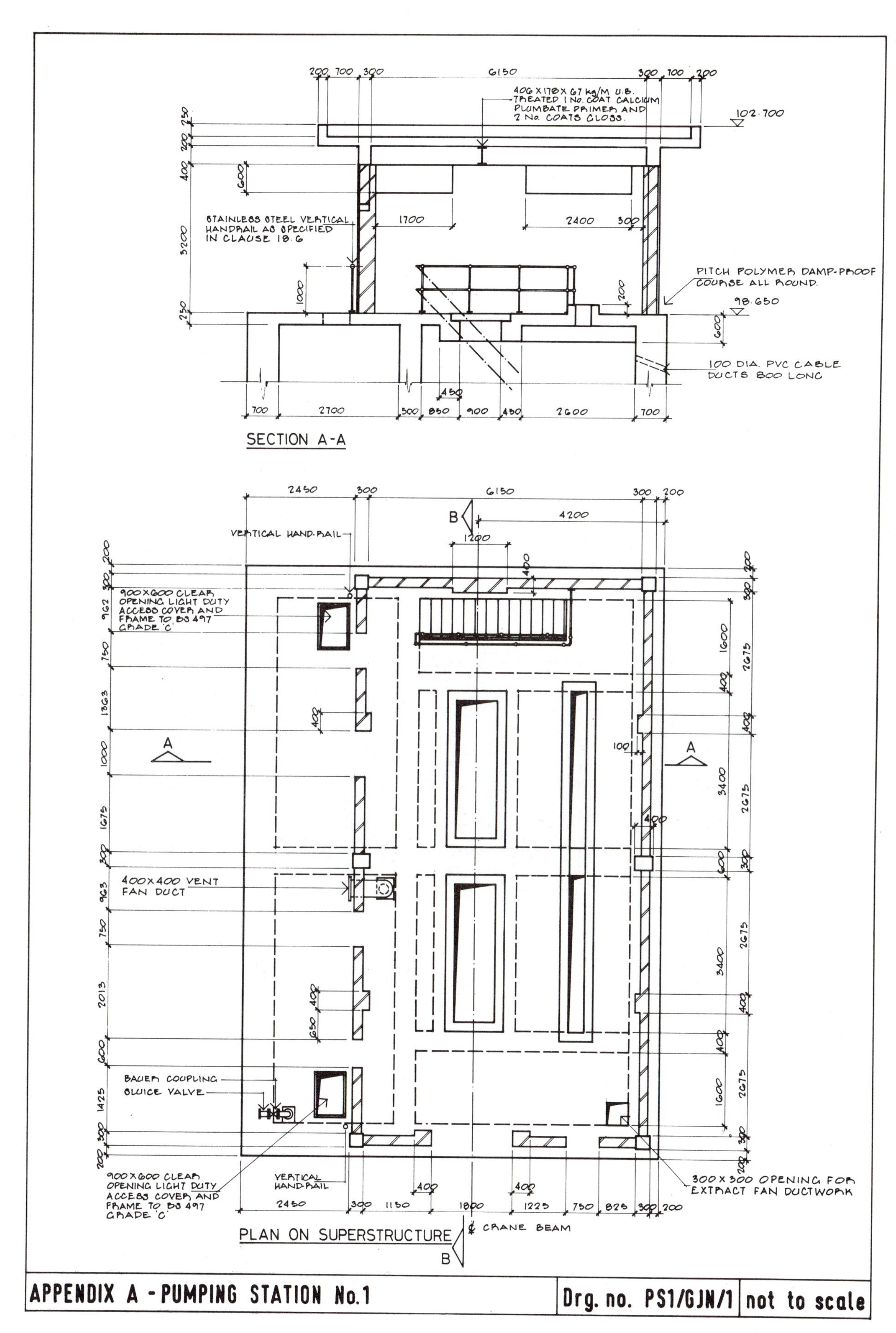

APPENDIX A - PUMPING STATION No.1 | **Drg. no. PS1/GJN/1** | **not to scale**

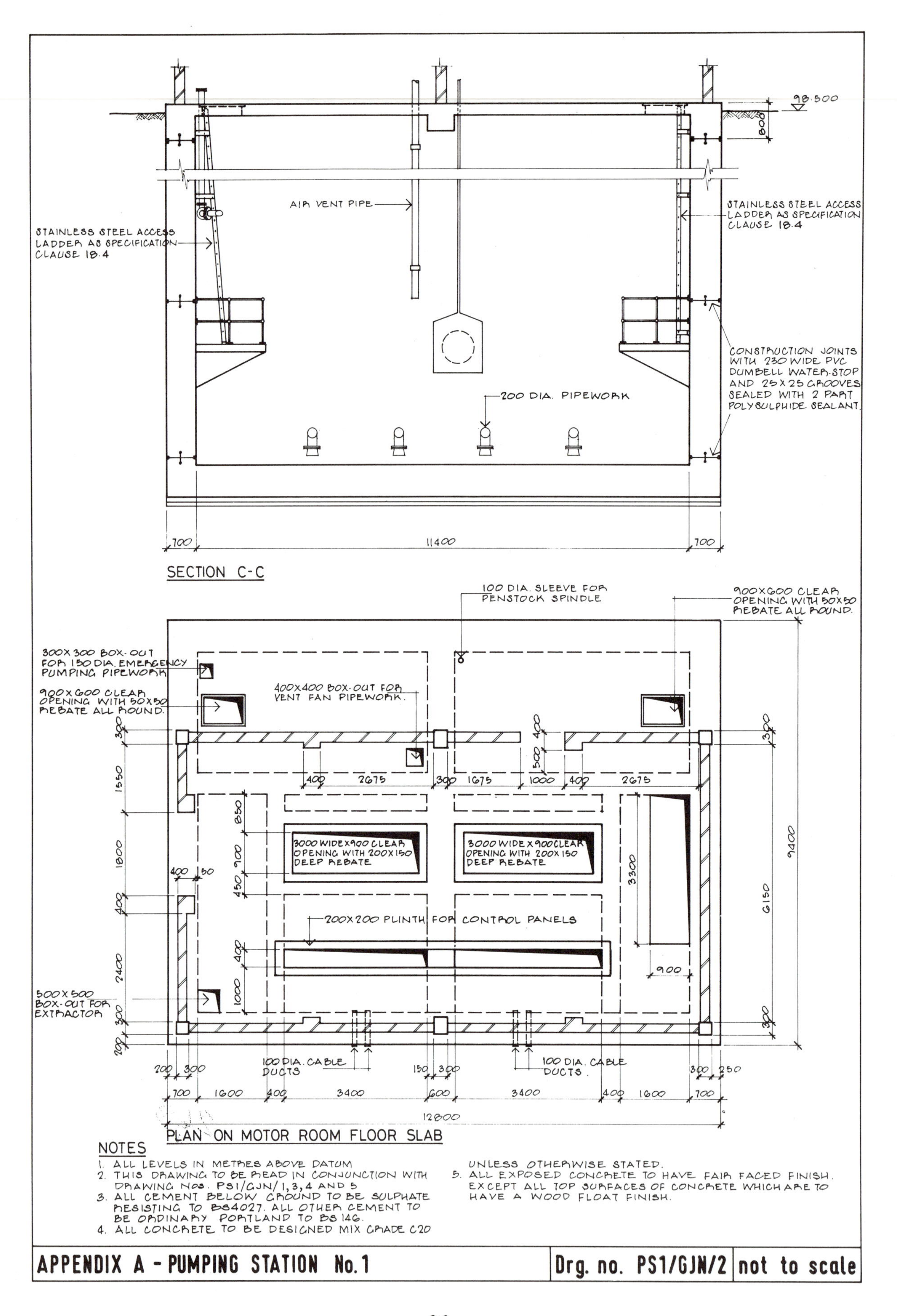
98.500
800
AIR VENT PIPE
STAINLESS STEEL ACCESS LADDER AS SPECIFICATION CLAUSE 18.4
STAINLESS STEEL ACCESS LADDER AS SPECIFICATION CLAUSE 18.4
CONSTRUCTION JOINTS WITH 230 WIDE PVC DUMBELL WATER-STOP AND 25X25 GROOVES SEALED WITH 2 PART POLYSULPHIDE SEALANT.
200 DIA. PIPEWORK
700
11400
700
SECTION C-C
100 DIA. SLEEVE FOR PENSTOCK SPINDLE
900X600 CLEAR OPENING WITH 50X50 REBATE ALL ROUND.
300X300 BOX-OUT FOR 150 DIA. EMERGENCY PUMPING PIPEWORK
900X600 CLEAR OPENING WITH 50X50 REBATE ALL ROUND.
400X400 BOX-OUT FOR VENT FAN PIPEWORK.
3000 WIDE X 900 CLEAR OPENING WITH 200X150 DEEP REBATE
3000 WIDE X 900 CLEAR OPENING WITH 200X150 DEEP REBATE
200X200 PLINTH FOR CONTROL PANELS
500X500 BOX-OUT FOR EXTRACTOR
100 DIA. CABLE DUCTS
100 DIA. CABLE DUCTS
12800
9400
PLAN ON MOTOR ROOM FLOOR SLAB
NOTES
1. ALL LEVELS IN METRES ABOVE DATUM UNLESS OTHERWISE STATED.
2. THIS DRAWING TO BE READ IN CONJUNCTION WITH DRAWING Nos. PS1/GJN/1, 3, 4 AND 5
3. ALL CEMENT BELOW GROUND TO BE SULPHATE RESISTING TO BS4027. ALL OTHER CEMENT TO BE ORDINARY PORTLAND TO BS 146.
4. ALL CONCRETE TO BE DESIGNED MIX GRADE C20
5. ALL EXPOSED CONCRETE TO HAVE FAIR FACED FINISH. EXCEPT ALL TOP SURFACES OF CONCRETE WHICH ARE TO HAVE A WOOD FLOAT FINISH.
APPENDIX A - PUMPING STATION No.1
Drg. no. PS1/GJN/2
not to scale

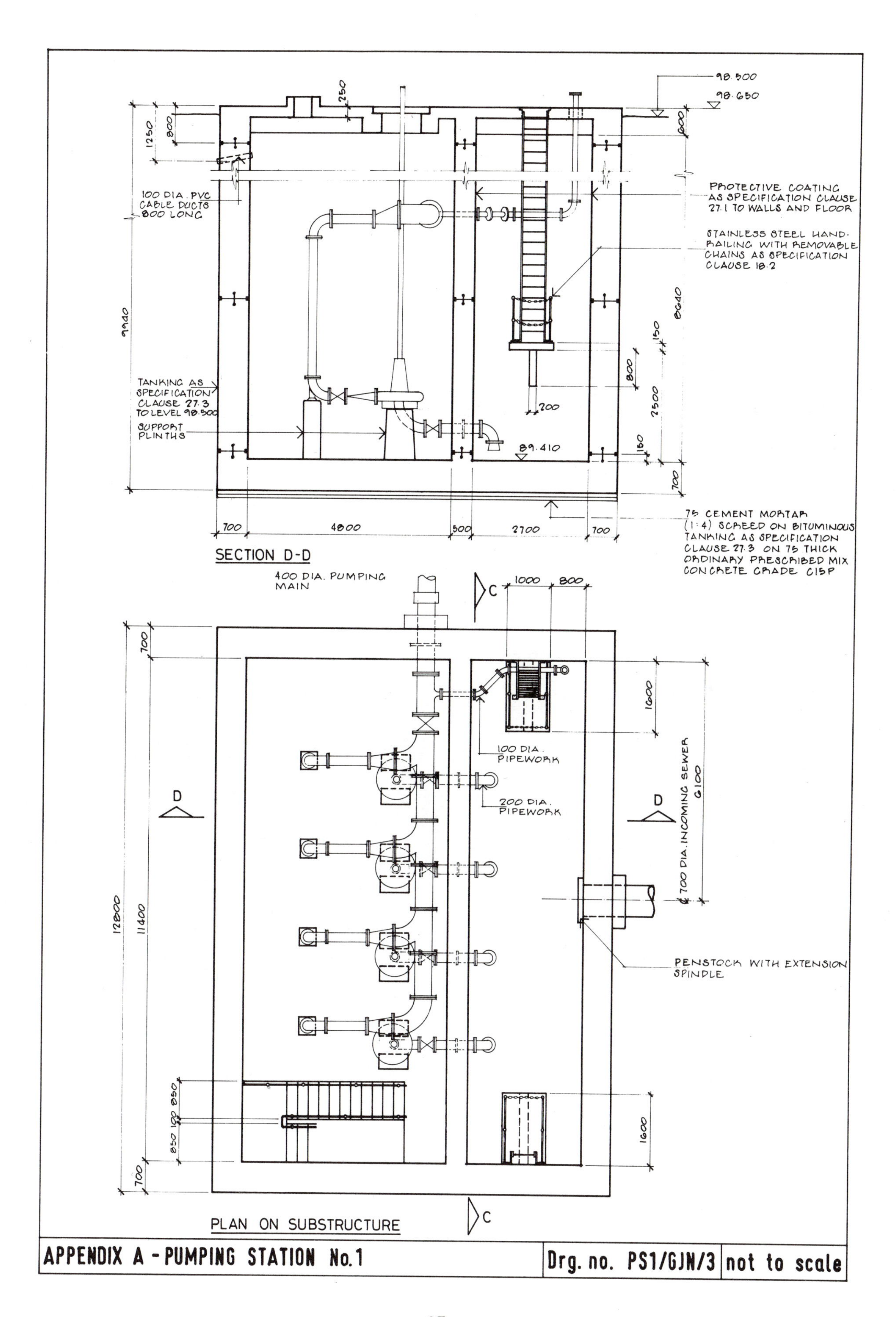

APPENDIX A - PUMPING STATION No.1 | Drg. no. PS1/GJN/3 | not to scale

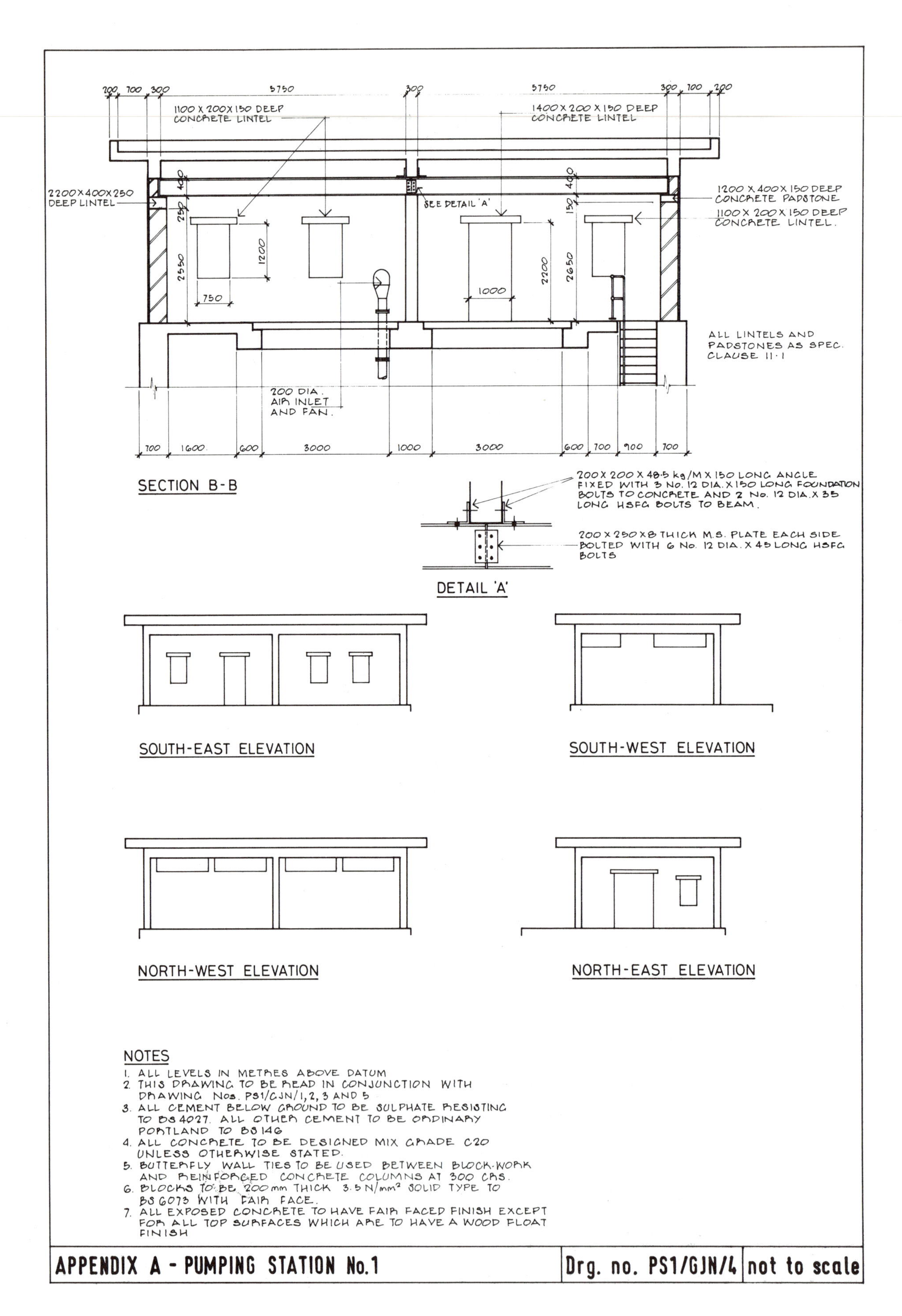

1100 X 200 X 150 DEEP CONCRETE LINTEL
1400 X 200 X 150 DEEP CONCRETE LINTEL
2200 X 400 X 250 DEEP LINTEL
SEE DETAIL 'A'
1200 X 400 X 150 DEEP CONCRETE PADSTONE
1100 X 200 X 150 DEEP CONCRETE LINTEL.
ALL LINTELS AND PADSTONES AS SPEC. CLAUSE 11·1
200 DIA. AIR INLET AND FAN.
SECTION B-B
200 X 200 X 48·5 kg/M X 150 LONG ANGLE FIXED WITH 3 No. 12 DIA. X 150 LONG FOUNDATION BOLTS TO CONCRETE AND 2 No. 12 DIA. X 35 LONG HSFG BOLTS TO BEAM.
200 X 250 X 8 THICK M.S. PLATE EACH SIDE BOLTED WITH 6 No. 12 DIA. X 45 LONG HSFG BOLTS
DETAIL 'A'
SOUTH-EAST ELEVATION
SOUTH-WEST ELEVATION
NORTH-WEST ELEVATION
NORTH-EAST ELEVATION
NOTES
1. ALL LEVELS IN METRES ABOVE DATUM
2. THIS DRAWING TO BE READ IN CONJUNCTION WITH DRAWING Nos. PS1/GJN/1, 2, 3 AND 5
3. ALL CEMENT BELOW GROUND TO BE SULPHATE RESISTING TO BS 4027. ALL OTHER CEMENT TO BE ORDINARY PORTLAND TO BS 146
4. ALL CONCRETE TO BE DESIGNED MIX GRADE C20 UNLESS OTHERWISE STATED.
5. BUTTERFLY WALL TIES TO BE USED BETWEEN BLOCK-WORK AND REINFORCED CONCRETE COLUMNS AT 300 CRS.
6. BLOCKS TO BE 200mm THICK 3.5 N/mm² SOLID TYPE TO BS 6073 WITH FAIR FACE.
7. ALL EXPOSED CONCRETE TO HAVE FAIR FACED FINISH EXCEPT FOR ALL TOP SURFACES WHICH ARE TO HAVE A WOOD FLOAT FINISH
APPENDIX A - PUMPING STATION No.1
Drg. no. PS1/GJN/4
not to scale

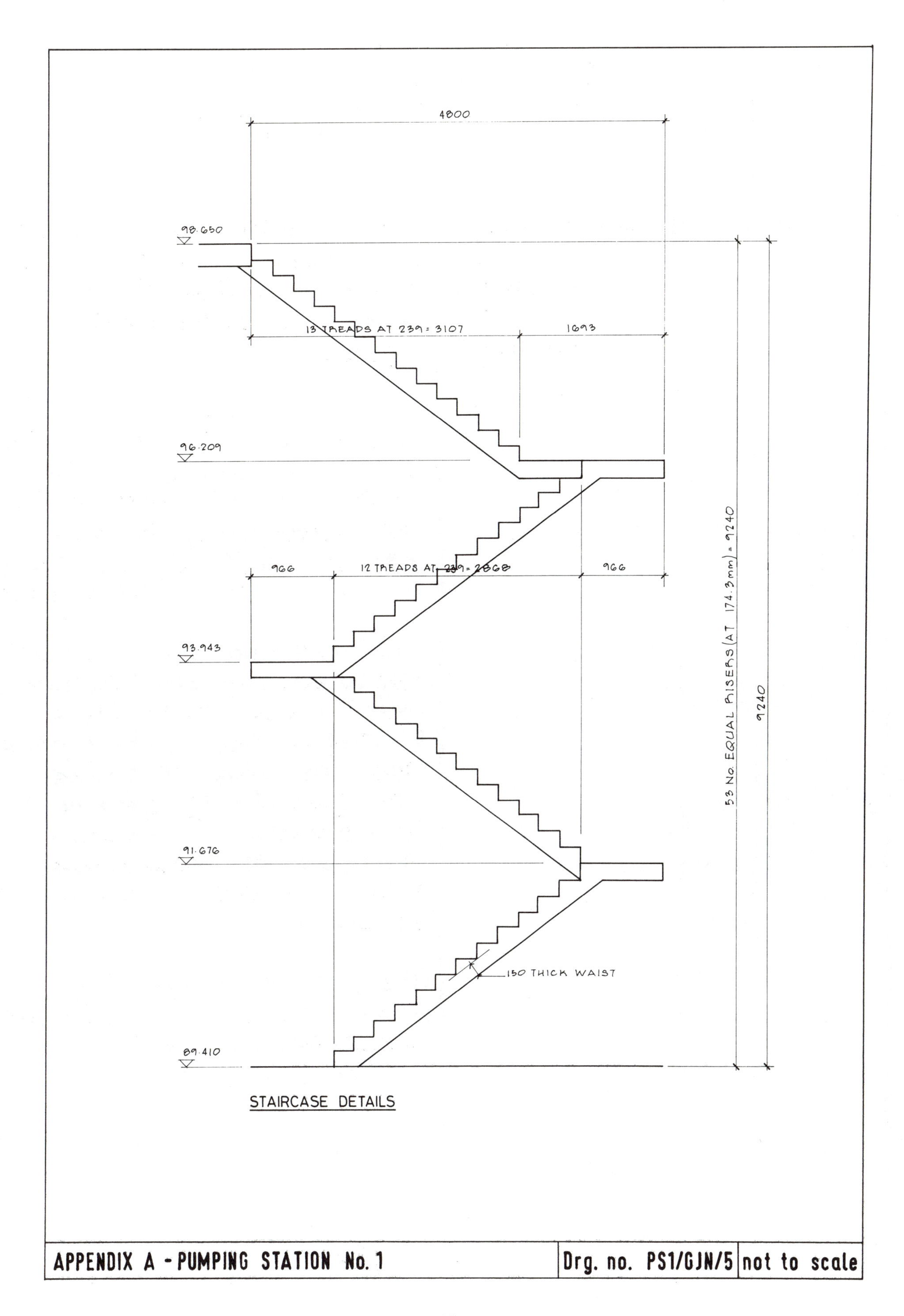
4800
98.650
13 TREADS AT 239 = 3107
1693
96.209
966
12 TREADS AT 239 = 2868
966
93.943
91.676
150 THICK WAIST
89.410
53 No. EQUAL RISERS (AT 174.3mm) = 9240
9240
STAIRCASE DETAILS
APPENDIX A - PUMPING STATION No. 1
Drg. no. PS1/GJN/5
not to scale

Pumping Station 1

				Explanatory Notes
	Drawing Numbers		PS1/GJN/1 PS1/GJN/2 PS1/GJN/3 PS1/GJN/4 PS1/GJN/5	The following example shows a typical reinforced concrete pumping station with a concrete and blockwork superstructure. It is assumed that the existing site is level with 200 mm of topsoil, with a rockhead at level 94.600. The surrounding finished ground level is also taken as being the existing level. This example demonstrates measurement in accordance with Classes E, F, G, H, N, U, V and W only. The pipework has not been measured, nor has the reinforcement to concrete. A separate example of reinforcement measurement is given elsewhere. The structure has been measured in two elements, substructure and superstructure.

Pumping Station 1

Dimensions	Squaring	Description	Explanatory Notes
		SUBSTRUCTURE	**Explanatory Notes**
		ETHWKS	This is the depth classification for all the materials in the excavation. This is because there is no express requirement to excavate the void in stages (M5), and the Commencing and Excavated Surface for all materials would be the same i.e. the top and bottom of the void.
		(depth for classfctn)	
		98500	
		89410	
	Slab 700 Screed 75 blindg 75	850 88560	
		9940	
		Excavn for foundns: Pumping Station 1	Although drawing PS1/GJN/2 shows the overall dimensions, the prudent taker off would check these by adding the individual dimensions. The volume of topsoil is based on the net plan area, as is all excavation. It is not necessary to state the Commencing Surface as it is also the Original Surface. The Excavated Surface is not stated because it is also the Final Surface (Paragraphs 1.12 and 1.13 and Rule A4)
9.40 12.80 0.20	24.06	Topsoil; max depth 5-10m (E 316) & Disposal (E 531)	None of the excavated material is required for re-use and there are no particular requirements for its disposal. It is therefore deemed to be taken off site (D4).

Pumping Station 1

			Excavn for foundns (cont)	Explanatory Notes
			Dep. 98500	
			topsoil 200	
			98300	
			rock level 94600	
			3700	
			Matl other than topsoil, rock or artfl hard matl, max depth 5-10m. (E326)	The depth of material is calculated by deducting the rock head level from the ground level after topsoil has been stripped.
	9.40			
	12.80			
	3.70		&	
		445.18	Disposal (E532)	
			Dep	
			rock level 94600	
			89410	
			base 700	
			blinding 75	
			screed 75 850 88560	
			6040	
			Rock; max depth 5-10m (E336)	The depth of rock is calculated from the top of the rock head to the underside of the blinding.
	9.40			
	12.80			
	6.04		&	
		726.73	Disposal (E533)	

Pumping Station 1

			Excavn. Anc.	Explanatory Notes
			Prepn of excavated surfs; rock (E523)	Preparation of surfaces is measured only to the underside of the blinding as this is the only surface to receive Permanent Works. The sides of the excavation will have formwork measured and no preparation is measured (M11). As no angle of inclination is stated, it is assumed that the preparation is less than 10° to the horizontal (A7)
	9.40 12.80	120.32		
			IN-SITU CONCRETE	
			Provn. of conc.	
			Standard mix ST3; sulphate resistg; cement to B.S. 4027; 20 agg. to B.S. 882 (F137)	The quantities for provision of concrete are calculated last, and are abstracted from the squared dimensions for the placing of concrete items.
	9.63	9.63m³	(blinding)	Because of the rounding up and down of dimensions, it is possible to have provision of concrete items which are slightly different to the total billed quantities for the placing items.
			Designed mix grade C20; sulphate resistg cement to B.S. 4027; 20 agg. to B.S. 882 (F247) (structure)	
	84.22			
	0.74			
Ddt 3.62	26.69			
	52.67			
Ddt 0.27	263.89			
	0.65			
	1.91			
	1.85			
		428.73m³		

Pumping Station 1

Timesing	Dimensions	Squaring	Description	Explanatory Notes
			Placg of mass conc.	
	9.40 12.80 0.08	9.63	Blinding; thness n.e. 150mm (F511)	It is not necessary to state that blinding is placed against the excavated surface (A2). The top surface of the blinding would have a general levelling and no further items would be measured under G81*
			Placg of r/f'd conc	
	9.40 12.80 0.70	84.22	Bases, footgs, pile caps and grnd slabs; thness ex. 500mm (F624)	Note that up to this point the annotation or 'signposting' of dimensions has been minimal. This is because it has been quite obvious where they apply i.e there is only one blinding layer and only one base.
2/	1.00 1.60 0.15	0.48	Susp. slabs; thness n.e. 150mm (F631)	The supporting concrete to the platform is an integral part of the slab and therefore classed and measured as part of it.
2/½/	1.60 0.20 0.80	0.26 0.74	(platform at bottom of ladders)	

Pumping Station 1

			Placg of r/f'd conc (cont)	Explanatory Notes
			Susp. slabs; thness 150-300mm (F632)	The suspended slabs are measured from the inside face of the walls. The classification of the thickness ignores the presence of the attached beams and upstands (D7)
	4.80 11.40 0.25	13.68	(roof (pump chamber)	
	2.70 11.40 0.25	7.70	(penstock chamber)	
	2.70 0.60 0.35	0.57	(beams (penstock chamber)	Beams integral with suspended slabs are classed as suspended slabs (M4). In this case the upstand to the control panels is measured separately. Beams are measured to the inside face of the walls from which they span.
2/	4.80 0.40 0.35	1.34		
	4.80 0.60 0.35	1.01	(pumpchamber across width)	
2/2/	0.90 0.20 0.35	0.25		
2/2/	3.40 0.45 0.35	2.14	(pump chamber spanning between previous)	
		26.69		

Pumping Station 1

			Placg of r/f'd conc (cont)	Explanatory Notes
			Ddt	
			Susp. slabs; thness 150-300mm abd (F632)	These openings must be deducted from the concrete volume as their area exceeds 0.5 m². All other openings in the slab are measured as voids and are therefore not deducted (see D3 of Class G)
2/	0.90 3.00 0.10	0.54	(rebated openings) dep 600; 150 / 350, 500, 100	
2/	1.30 3.40 0.15	1.33		
	0.40 7.40 0.25	0.74	(control panel opening)	
	0.90 3.30 0.25	0.74	(staircase opening)	
2/	0.90 0.60 0.25	0.27	(access covers above penstock chamber)	
		3.62		
			height 98650 89410 9240	
			Add	
			Walls; thness 300-500mm (F643	The walls are measured from the top of the base slab to the top of the suspended slab. No deductions are made for joint components or cast in components unless they exceed 0.1m³ (M1).
	11.40 0.50 9.24		(int wall)	
		52.67		
			Walls; thness ex 500mm (F644)	
2/	9.00 0.70 9.24	116.42	(extl walls)	
2/	11.40 0.70 9.24	147.47		
		263.89		

Pumping Station 1

Timesing	Dimensions	Squaring	Description
			<u>Placg. of r/f'd conc (cont')</u>
			<u>Ddt</u>
			Walls; thness ex. 500mm abd (F644)
π/	0.35		
	0.35		
	0.70	0.27	
			<u>Add</u>
			Upstands; size 200 x 200mm (F680)
2/	7.40		
	0.20		
	0.20	0.59	(control panel)
2/	0.80		
	0.20		
	0.20	0.06	
		0.65	
			Stairs and landings (F680.1)
	1.69		
	0.85		
	0.18	0.26	(top landing)
	0.97		
	0.95		
	0.18	0.17	
2/	0.97		
	1.80		
	0.18	0.63	(intermediate landings)
3/12/½/	0.85		
	0.24		
	0.17	0.62	(bottom flights steps)
13/½/	0.85		
	0.24		
	0.17	0.23	(top flight steps)
		1.91	

Explanatory Notes

The incoming sewer is the only cast in component large enough to warrant a deduction from the volume.

Although these have been classified under other concrete forms, they could have been measured under the classification for walls.

Plan of top landing

1690

850

1800

966

The top step and associated waist of the lower three flights will have been measured with the landings. This is acceptable because both items are combined under the same description. (See next item for calculation of step and waist volumes).

Pumping Station 1

Timesing	Dimensions	Squaring	Description
			Placg r/f'd conc (cont)
			Stairs & landings (cont)
			Len
			$x^2 = 239^2 + 174.34^2$
			$x = 295.83$
			× 13
			3846
			295.83
			× 12
			3550
			(waists)
	3.85		
	0.85		
	0.15	0.49	(top flight)
3/	3.55		
	0.85	1.36	(bottom flights)
	0.15		
		1.85	
			CONCRETE ANCILLARIES
			Fmwk. Rough Fin
			dep.
			Top joint 98650 + 800 = 97850
			bottom joint 89410 + 150 = 89560
			8290
			Plane vert. EX. 1.22m (G145)
2/	9.40		
	8.29	155.85	(extl wall outside face)
2/	12.80		
	8.29	212.22	
		368.07	

Explanatory Notes

The calculation of the length of the waist is done by calculating the length for each individual step and multiplying it by the number of steps in the flight. The calculation of the volume of the steps is done by calculating the cross-sectional area of the step and multiplying by its length.

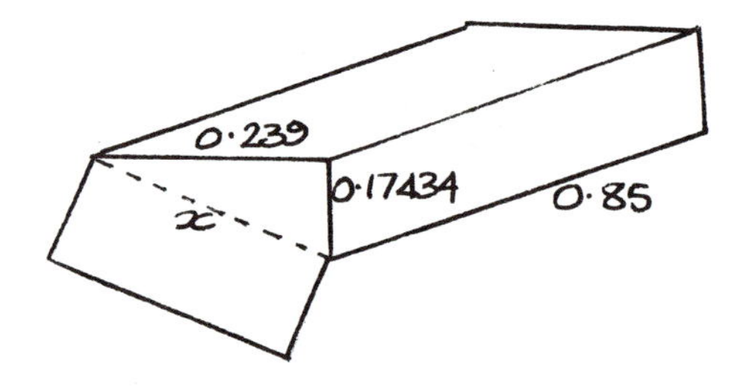

Formwork to the top and bottom of the wall is classed as width 0.4–1.22 because of the the position of the joints and is therefore excluded from this measurement.

Pumping Station 1

Timesing	Dimension	Squaring	Description	Explanatory Notes
			Fmwk. Rough. Fin (cont)	
			dep: top joint 800; fair face 150; 650	
			Plane vert 0.4–1.22m (G144) (extl wall outside face)	The formwork to the areas above and below the top and bottom joints are classed as 0.4–1.22m. The upper 150mm of the top joint formwork is exposed to view and must be fair faced. It is deducted from the total area and measured separately, classed as 0.1–0.2m wide.
2/	9.40 0.65	12.22	(top joint)	
2/	12.80 0.65	16.64		
2/	9.40 0.85	15.98	(bottom joint)	
2/	12.80 0.85	21.76		
		66.60		
			Fmwk. Fair Fin	
			Plane vert. 0.1–0.2m (G242)	The formwork to the inside face of the walls is classed as 0.1–0.2m at the base because of the presence of the kicker joint.
2/	9.40	18.80	(top 150mm of outside face of extl wall)	
2/	12.80	25.60		
2/	4.80	9.60	(internally at kicker level to all walls)	
2/	2.70	5.40		
2/2/	11.40	45.60		
		105.00		
			Plane vert ex. 1.22m (G245) (internally between bottom and top joint to all walls)	
2/	4.80 8.29	79.58		
2/	2.70 8.29	44.77		
2/2/	11.40 8.29	378.02		
		502.37		

Pumping Station 1

			Fmwk. Fair Fin (cont')	Explanatory Notes
			dep 800 256 550	
2/	4.80 0.55	5.28	Plane vert. 0.4–1.22m (G 244)	It is recommended that no deduction is made from the area of formwork
2/	2.70 0.55	2.97	(internally between top joint and U/S of slab)	between the top joint and the underside of the slab for
2/2/	11.40 0.55	25.08		intersections with beams because their cross-sectional area does
		33.33		not exceed $0.5m^2$
			Len 1600 400 3400 5400	
2/	2.70 5.40	29.16	Plane horiz. ex 1.22m (G 215) (penstock chamber roof)	Narrow width formwork caused by openings is only classed as such where formwork to the
Ddt 2/ 0.90 0.60		1.08	(for access covers)	openings is measured under the general formwork classification and not as voids as defined in
	1.60 4.80	7.68	(pump chamber roof)	D.3. The only areas to the soffit of the pump chamber
	1.60 1.50	2.40		roof which are not below 1.22m wide are the area at the opposite
		38.16		end of the stairs and the area between the staircase opening and the long external wall (see diag.)

Pumping Station 1

Timesing	Dimensions	Squaring	Description
			Fmwk. Fair. Fin (cont')
			Ddt Plane horiz. 0.4-1.22m
	0.70 3.30	2.31	(next stairs) (G214) (soffit of pump chamber
2/	3.40 1.00	6.80	(either side of plinth panel opening)
2/	3.40 1.20	8.16	
2/	3.40 0.45	3.06	(between intl wall & rebated opening)
		20.33	
			Add Plane. horiz. 0.4-1.22m
	4.80 0.60	2.88	(central beam across width of pump chamber) (G214) (beam soffits
2/	0.90 0.20	0.36	
2/	0.90 0.60	1.08	(outside beams across width of pump chamber)
2/	3.40 0.45	3.06	(long beams to rebated openings)
		7.38	
			Plane horiz. 0.2-0.4m
2/	4.80 0.40	3.84	(G213) (outside beams across width of pump chamber)
Ddt 2/ 0.90 0.40	0.72		
		3.12	

Explanatory Notes

Because of the smaller nature of the work, formwork to beams is measured in accordance with G1-4 1-4* and not G1-4 8*

The beams which span the width of the pump chamber widen out where they cross the rebated beams. In the case of the central beam this simply means an additional area of formwork in the 0.4-1.22 classification. The outside beams however are classified as 0.2-0.4m and this widening means that the beam is classified as 0.4-1.22m for the length of the additional width

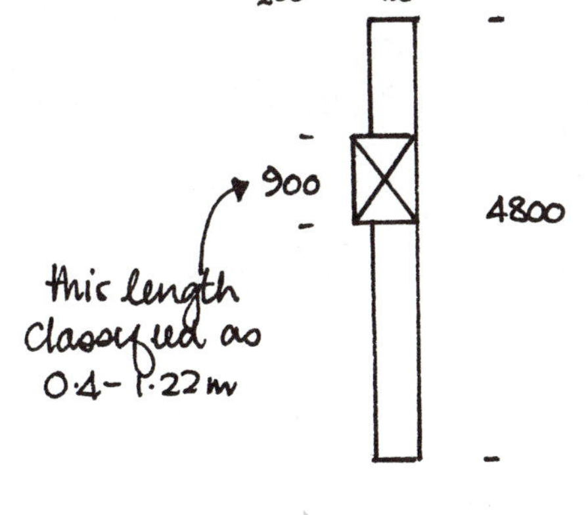

Pumping Station 1

Timesing	Dimensions	Squared	Description	Explanatory Notes
			Fmwk. Fair Fin (cont)	
			Plane vert. 0.2-0.4m (G243)	The sides of the long beams on either side of the rebated openings are in two different width classifications
2/	2.70 0.35	1.89	(penstock chamber) (beam sides	
2/	4.80 0.35	3.36	(outside beams across width of pump chamber)	
2/	2.60 0.35	1.82		
2/	0.40 0.35	0.28		
2/2/	3.40 0.35	4.76	(long beams to rebated opening -outside face only)	
		12.11		
			Plane vert. 0.4-1.22m (G244)	These holes are above the minimum size to enable measurement under the voids classification and have to be measured under the general formwork classification. Their area must therefore be deducted
2/2/	3.00 0.45	5.40	(rebated openings) (opngs in roof slab	
2/	0.90 0.45	0.81		
2/	7.40 0.45	6.66	(control panel openings - long side)	
		12.87		
			Plane. vert. 0.2-0.4m (G243)	
2/	0.40 0.45	0.36	(control panel opng - short sides) (opngs in roof slab	
2/2/	0.90 0.25	0.90	(access covers in penstock chamber)	
2/2/	0.60 0.25	0.60		
	3.30 0.25	0.83	(staircase)	
	0.90 0.25	0.23		
		2.92		

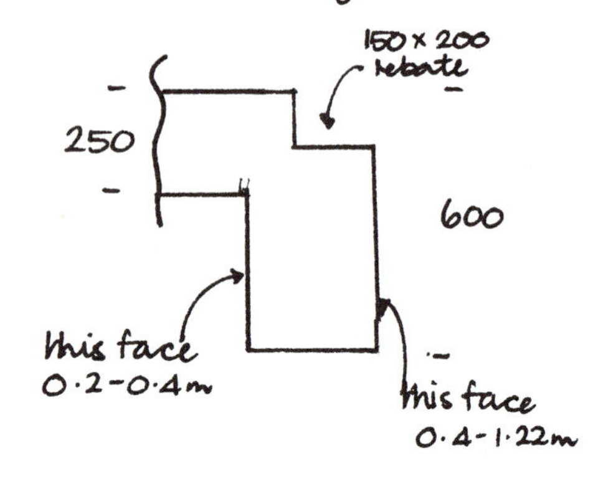

Pumping Station 1

Timesing	Dimensions	Squaring	Description	Explanatory Notes
			Fmwk. Fair Fin (cont)	
			Plane vert. 0.1-0.2m (G242)	
2/2/	3.40	13.60	(rebates to rebated opngs) (opngs in roof slab	
2/2/	1.30	5.20		
2/	8.00	16.00	(control panel opng - side of plinth)	
2/	0.80	1.60		
		36.40		
			Plane horiz; 0.2-0.4m (G213)	
2/2/	0.40 1.60		(landings to ladders)	
		2.56		
			Plane vert. 0.1-0.2m (G242)	
2/2/	1.60	6.40	(landings to ladders)	
2/	1.00	2.00		
		8.40		
			Plane vert; average width 0.4m (G246)	
2/2/½/	0.80 1.60		(landings to ladders)	
		2.56		
			Plane slopg; 0.1-0.2m (G222)	
			$x^2 = 1.6^2 + 0.8^2$ $x = 1.78$	
2/	1.78			
		3.56	(landings to ladders.	

Strictly speaking the formwork to the sides of the support to the landing should be classified as above. With such small quantities this would be unhelpful and therefore an additional description in accordance with Paragraph 5.10 is incorporated into the description and the average width is stated.

Pumping Station 1

			Fmwk. Fair Fin (cont')	Explanatory Notes
			Plane vert. 0.1–0.2m (G 242) (Stairs	Because of the 100mm gap between the stairs, the top riser in the bottom three flights will effectively be 100mm longer
14/	0.85	11.90	(risers-top flight)	
3/	0.95	2.85	(risers-bottom flight)	
3/12/	0.85	30.60	Len 1693 966 727	
	0.73	1.69	(top landing)	
	0.96			
2/	0.96	1.92	(intermediate landing)	
		48.96		
			Plane vert. 0.2–0.4m (G 243) (sides of stairs	The classification for the width of formwork to the waist is taken as the maximum width and the area is taken as the gross area. The top flight and flight second from bottom have formwork measured to one side only as the other side is cast against the external wall. The calculation of the area is done by adding the area of the waist to twice the area of the side of the step.
	3.85 0.15	0.58	(top flight)	
13/½/	0.24 0.17	0.27		
2/2/	3.55 0.15	2.13		
2/12/½/	0.24 0.17	0.49	(bottom flight)	
	3.55 0.15	0.53		
12/½/	0.24 0.17	0.24		
		4.24		
			Plane slopg 0.4–1.22m (G 224) (Stairs	
	3.85 0.85	3.27	(soffits)	
3/	3.55 0.85	9.05		
		12.32		

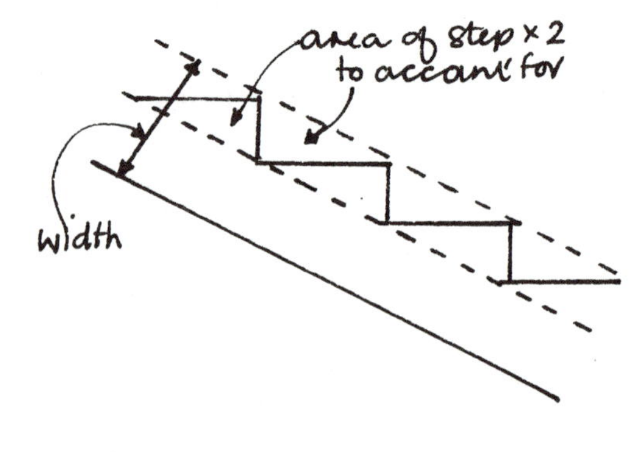

Pumping Station 1

Timesing	Dimensions	Squaring	Description	Explanatory Notes
			Fmwk. Rough Fin.	**Explanatory Notes**
			For voids	It is assumed that the formwork to the sides of the voids will have a rough finish because they will be eventually filled in around the insert.
			small void depth n.e. 0·5 (G171)	
	1		(emergency pumping pipework) (in roof	
		1		
			large void depth n.e. 0·5 (G175)	
	1		(vent fan)	
	1		(extractor) (in roof.	
		2		
			Joints	
			Open surf plain 0·5 – 1m (G612)	The average width of the joint is measured from outside face to outside face with no deduction for the width occupied by the rebate (M11)
2/3/	9.00 0·70	37·80	(jnts at all three levels	There is no surface treatment to the joints so they are classed as plain.
2/3/	11·40 0·70	47·88	extl walls)	
		85·68		
			Open surf plain n.e. 0·5m (G611)	
3/	11·40 0·50		(int wall)	
		17·10		

Pumping Station 1

Timesing	Dimensions	Squared	Description	Explanatory Notes
			Joints (cont'd)	**Explanatory Notes**
			PVC waterstop; dumbell type, 230 wide (G653)	As the width of the waterstop is stated it is not necessary to also state the average width in the ranges given in the Third Division. Note that there is no separate measurement for angles as these are deemed to be included (C4)
3/	36.00	108.00	(extl walls)	
3/	12.10	36.30	(int walls)	
		144.30		
			girth: 11400 + 8000 = 19400; x2 = 38800; less 4/2/½/700 = 2800; = 36000	
			len: 11.400; Ends 2/½/700 = .700; = 12.100	
			Sealed rebate or groove; 25 x 25 mm with two part polysulphide sealant (G670)	Although not specifically mentioned, the measurement of the groove is taken at its exposed face. No formwork is measurable to any work classed as joints (C3)
2/3/	36.00	216.00	(extl walls)	
2/3/	Ddt 0.70	4.20		
2/3/	11.40	68.40	(intl walls)	
		280.20		
			Conc. Accessories	
			Fin top surfs; wood float (G811)	No deduction is made for the areas of the support plinths as they do not exceed 0.5m² (M14)
	11.40 2.70	30.78	(penstock chamber floor)	
	11.40 4.80	54.72	(pump chamber floor)	
2/	1.00 1.60	3.20	(ladder landings)	
		88.70		

Pumping Station 1

Timesing	Dimension	Squaring	Description	Explanatory Notes
			Conc. Ascessories (Cont)	
			Fin top surfs; wood float (G811)	
13/	0.24 0.85	2.65	(treads) (stairs	
3/12/	0.24 0.85	7.34		
	1.69 0.85	1.44	(top landing)	
	0.97 0.95	0.92		
2/	0.97 1.80	3.49	(intermediate landings)	
		15.84		
			Fin. top surfs; wood float (G811)	
	12.80 9.40		(roof slab	
		120.32		
			Ddt	
			Fin. top surfs; Wood float (G811)	Although not clear from CESMM no deduction is made from the area for that which is subsequently covered by the brick walls. No deduction is made for openings less than $0.5m^2$ (M14)
	0.90 3.30	2.97	(stairs opng)	
2/	3.00 0.90	5.40	(rebated opngs)	
	7.40 0.40	2.96	(control panel opng)	
2/	0.90 0.60	1.08	(access covers)	
		12.41		

Pumping Station 1

			Inserts	Explanatory Notes
4/	1	4	100mm dia. PVC cable ducts, 800 long; proj. two sides (G832) (pump chamber extl wall at high level)	This item would also include for the supply of the duct as it has not been otherwise stated (C7)
	1	1	700mm dia. pipe proj. one side (supply incl. elsewhere) (G832.1) (sewer thro extl wall)	This is classified as projecting from one surface because the end of the pipe is flush with the internal face of the wall.
	1	1	400mm dia. pipe proj. two sides (supply incl. elsewhere) (G832.2) (pumpg main thro extl wall)	
4/	1	4	200 mm dia. pipe proj. two sides (supply incl. elsewhere (G832.3) (pumpg main thro intl wall)	
2/	1	2	100mm dia. pipe proj. two sides (supply incl. elsewhere) (G832.4) (emergency pumpg pipe thro intl wall) (sleeve for penstock spindle)	

Pumping Station 1

			Inserts (cont'd)	Explanatory Notes
			150mm dia pipe in 300 × 300 × 250mm deep preformed boxout proj. two sides as Spec Clause 12.1 (G 832.5)	Inserts grouted into preformed openings must be described. The works in grouting into preformed openings can be quite involved so reference is made to the relevant Specification Clause rather than give a detailed description
	1		(emergency pumpg pipe thro roof)	
		1		
			200mm dia pipe in 400 × 400 × 250mm deep preformed boxout proj. two sides as Spec Clause 12.1 (G832.6)	
	1		(Air inlet thro roof)	
		1		
			300 × 300mm duct in 500 × 500 × 250mm deep preformed box-out proj two sides as Spec Clause 12.1 (G832.7)	
	1		(Extractor fan thro roof)	
		1		

Pumping Station 1

Timesing	Dimension	Squaring	Description	Explanatory Notes
			Miscellaneous Metalwork	**Explanatory Notes**
			Ladders, stainless steel as Spec. Clause 18.4 and drwg no. PS1/GJN/2; incl. all fixing to conc. (N130)	Alternatively the ladders could have been enumerated stating the length. The item includes for all fixing to concrete (C1). Alternatively the pockets and bolts could have been measured in Class G.
2/	6.59	13.18	Len: 98650 − 89410 = 9240; 2500 + 150 = 2650; 9240 − 2650 = 6590	
			Handrails; level; stainless steel as Spec. Clause 18.2; incl. all fixing (N140)	Although not a specific requirement, level handrails and raking handrails are kept separate because it is considered that they have different cost considerations. Alternate flights will have a handrail either one or both sides depending on whether it is next to the external wall or not.
2/2/	1.60	6.40	(ladder landings)	
	8.53	8.53	(stair landings)	
		14.93	Len: 1200; top of stairs 3300; − 200; 727; 1st inter down 200; 966; − 2 & 3rd inters 966; 966; total 8525	
			Ditto sloping (N140.1)	x is the length of the sloping handrail which is the hypotenuse of a right angled triangle (Pythagoras' theorem)
	3.95		$x^2 = 3.10^2 + 2.45^2$; $x = 3.95$	
2/2/	3.66			
	3.66		$x^2 = 2.87^2 + 2.28^2$; $x = 3.66$	
		22.25		

Pumping Station 1

			Miscellaneous Metalwork (cont)	Explanatory Notes
	11·40	11·40	Handrail; level; stainless steel as Spec. Clause 18·6, incl. all fixing (N140·2) (extl)	Although this handrail and part of the handrail before is strictly speaking a superstructure item, they have been included here in order to keep all similar items of work together.
2/	1	2	Access covers & fr. 900 × 600 mm light duty to B.S. 497 grade C (N190·1) (penstock chamber roof)	There is no separate classification for access covers so this is a rogue item.
2/	1	2	Safety chains 900 mm long; stainless steel as Spec. Clause 18.2 incl. all fixing. (N190)	There is no separate classification for safety chains so this is a rogue item.

Pumping Station 1

			Waterproofing	Explanatory Notes
			Bituminous Tanking as Spec Clause 27.3	Because the description contains reference to the relevant specification clause, it is not necessary to go into detail about the number of coats and thickness of material.
			Upper surfs. n.e. 30° to horiz (W241)	
	12.80		(U/s of structure)	
	9.40			
		120.32		
			Surfs. ex. 60° (W243)	
2/	12.80		(outside of Extl. walls.)	
	9.87	252.67		
2/	9.40		ht: 98500, 89410, 9090; base 700; screed 75; 9865	
	9.87	185.56		
		438.23		
			Dampproofing	
			Protective coating as Spec. Clause 27.1 n.e 30° to horiz (W141)	
	2.70		(floor penstock chamber)	
	11.40			
		30.78		

Pumping Station 1

			Damp proofing (cont'd)	Explanatory Notes
			Protective coating as Spec. Clause 27.1 ex. 60° to horiz.	No deduction is made where the beams meet the walls or the sewer as the cross-sectional area does not exceed $0.5m^2$
2/	2.70 8.99	48.55	(W 143) (penstock chamber	
2/	11.40 8.99	204.97		
		253.52		
			Protective layers	
			Sand and cement (1:4) screed 75mm thick in one coat n.e. 30° to horiz.	
	12.80 9.40		(U/S of structure) (W441)	
		120.32		

Pumping Station 1

Timesing	Dimensions	Squaring	Description	Explanatory Notes
			SUPERSTRUCTURE	**Explanatory Notes**
			Provn. of conc.	
	1.30 0.86 4.24 24.28 2.24	32.92m³	Designed mix grade C20; OPC to B.S. 146; 20agg to B.S. 882 (F243)	
			Placg of r/f'd conc.	
4/	0.30 0.30 3.60	1.30	Columns and piers; X-sect. area 0.03 – 1m² (F652) 3200 400 3600	The columns are measured from the top surface of the pumphouse floor to the underside of the roof slab. The beams are then measured between the columns.
2/	0.30 0.40 3.60	0.86	Ditto 0.1 – 0.25m² (F653)	
2/	6.15 0.30 0.40	1.48	Beams; X-sect area 0.1 – 0.25m² (F663)	Again note that there is no annotation or 'signposting' of the dimensions as it is quite apparent on the drawing where the descriptions are derived from
2/2/	5.75 0.30 0.40	2.76		
		4.24		

Pumping Station 1

			Placg. of r/f'd conc. (cont')	Explanatory Notes
			Len 2/200 400 2/700 1400 2/300 600 2/5750 11500 300 14200	
			Wid 2/200 400 2/700 1400 2/300 600 6150 8550	
	14.20 8.55 0.20	24.28	Susp. slabs thness 150-300mm (F632) (roof)	The suspended slab is measured over the beams and under the perimeter kerb
	44.70 0.20 0.25	2.24	Upstands; size 200x 250mm (F680) (roof) girth 14200 8550 22750 x2 45500 4/2/½/200 800 44700	Again the perimeter kerb to the roof has been classified as an upstand rather than a wall as it is felt that this more accurately reflects the work involved.

Pumping Station 1

Timesing	Dimension	Squaring	Description	Explanatory Notes
			CONCRETE ANCILLARIES	
			Fmwk. Fair Fin.	
			To components of constant x-sect; columns size 300 x 300mm; extl walls of superst. (G282)	The measurement of the columns and beams to the superstructure is done as components of constant cross-section in order to demonstrate the classification. In practice on a contract of this size it is not of much benefit to the contractor in that the number of uses of the formwork will be very small.
4/	3·60	14·40		
			Ditto 300 x 400mm (G282·1)	
2/	3·60	7·20		
			To components of constant x sect; beams size 300 x 400mm; roof slab (G281)	To be strictly correct, the area of the columns in contact with the brickwork should be deducted from the area of the fair finish and added back as rough because it is not exposed to view. However, it is felt that this is totally impracticable and the fair finish is carried through.
2/	6·15	12·30		
2/2/	5·75	23·00		
		35·30		
			Plane horiz; 0·4-1·22m (G214)	
2/	14·20 0·90	25·56	(perimeter o'hang of roof)	
2/	6·75 0·90	12·15		
		37·71		

Pumping Station 1

Timesing	Dimensions	Squaring	Description
			Fmwk. Fair Fin. (Cont'd)
			Plane. horiz; ex. 1·22m (soff. of roof intl.) (G215)
2/	5·75		
	6·15		
		70·73	
			Plane vert. 0·1 - 0·2m (intl. face of upstand) (G242)
2/	13·80	27·60	
2/	8·15	16·30	
		43·90	
			Plane vert. 0·2 - 0·4m (extl. face of upstand) (G 243)
2/	14·20		
	0·45	12·78	
2/	8·55		
	0·45	7·70	
		20·48	
			Concrete Accessories
			Fin. top surfs; wood float (top of roof slab) (G.811)
	14·20		
	8·55		
		121·41	

Explanatory Notes

Pumping Station 1

Timesing	Dimension	Squaring	Description	Explanatory Notes
			Concrete Accessories (cont)	
2/6/	3.20	38.40	Butterfly wall ties; 300mm centres; proj from one surf (G831) (betw. column & block wall)	Alternatively the measurement of the wall ties could have been included in Class U, but it would still have been necessary to measure an item in Class G for the building in. Although the ties have been measured under the linear inserts classification stating the spacing, they could have been enumerated.
2/3/	1	6	fndn. bolts 12 dia x 150mm long; proj from one surf (G832) (fixing crane rail to central conc. beam)	
			PRECAST CONCRETE	
			Lintels	There is no separate classification for lintels in Class C and this is therefore a rogue item. Because the actual size is stated it is considered that it is not necessary to also state the weight. The specification clause would contain details of reinforcement etc., or alternatively this may be contained on a drawing
3/	1	3	1100 x 200 x 150mm deep as Spec. Clause 11.1 (H900) (window)	
	1	1	1400 x 200 x 150mm deep as Spec. Clause 11.1 (personnel door) (H900.1)	

Pumping Station 1

			Description	Explanatory Notes
			Lintels (cont)	
			2200 × 400 × 150mm deep as Spec. Clause 11.1 (H900.2)	
	1	1	(door under crane beam)	
			Padstones	
			1200 × 400 × 150mm deep as Spec. Clause 11.1 (H900.3)	
	1	1		
			STRUCTURAL METALWORK	
			Fabrication of crane beams; straight on plan, 406 × 178 × 67 kg/m U.B. (M421)	The weight of the plate and angle must also be included in the mass of steel measured, but not the bolts (M3). The section size and weight is also given to assist the Contractor although this is not strictly necessary.
	11.80 0.067	0.79	(UB)	
2/	0.15 0.049	0.02	(angles) Len 5750 300 5750 = 11800	
2/	0.25 0.25 0.06	0.01	(plate)	
			&	
		0.82 kg ÷1000		
		0.01t	Permanent Erection (M720)	

Pumping Station 1

Timesing	Dimensions	Squaring	Site bolts	Explanatory Notes
			HSFG general grade	Although not a specific requirement the actual diameter and length of bolts is also stated. The bolts which fix the angles to the concrete are measured in Class G.
6/	1		12 dia x 45mm long (plate) (M741)	
		6		
2/2/	1		12 dia x 35mm long (flange/angle) (M741.1)	
		4		
			BRICKWORK AND BLOCKWORK	
			Dense concrete blocks to BS 6073 part 1 3.5N/mm² solid; 450 x 225 x 200mm; jointing in cement mortar (1:3); Stretcher bond.	
2/2/	5.75 3.20	73.60	Vert, straight walls; 200mm th. (U521) (extl. walls)	The measurement of the external walls is taken over all openings and holes for which the deductions are made separately.
2/	6.15 3.20	39.36		
		112.96		
	1.20 2.65		Vert. straight walls; 400mm th. (U531) (pilaster to S.W. wall.)	
		3.18		

Pumping Station 1

Timesing	Dimensions	Squaring	Dense conc. blocks (cont')	Explanatory Notes
			Ddt Vert, straight walls 200mm th. (US21)	
	1.70 0.60	1.02	(for opngs) (SW Elev.	The length of the attached pilaster to the N.W wall exceeds 4 times its projecting width so that the area of wall is measured separately as 300 thick. The area of the padstone is less than $0.25 m^2$ in area and is not deducted (M2)
	2.40 0.60	1.44		
	1.20 2.65	3.18	(for attached pilaster)	
	0.75 1.20	0.90	(window) (NE elev.	
	1.80 2.55	4.59	(door)	
	2.20 0.40	0.88	(lintels to same	Although the smaller lintels are less than $0.25 m^2$ they are deducted from the overall area because they form part of the window openings and their combined area exceeds the minimum.
	1.10 0.20	0.22		
	1.00 2.20	2.20	(door) (SE elev	
2/	0.75 1.20	1.80	(windows)	
	0.60 0.75	0.45		
	1.40 0.20	0.28	(lintels to same)	
2/	1.10 0.20	0.44		
4/	2.40 0.60	5.76	(opgs) (NW Elev.	
		23.16		

Pumping Station 1

Timesing	Dimensions	Squaring	Surface features	Explanatory Notes
			Add Pilasters size 400×100mm	Although the cross-sectional area of the pilasters is less than that specified in A7, the size is still stated because it is considered that this gives the Contractor the information necessary to identify any cutting or special blocks in accordance with A.6.
2/2/	3·20	12·80	(to N.W. & S.W walls) (U576)	
2/	2·55	5·10	(either side of main door)	
		17·90		
			Fair facing; flush pointg as work proceeds (U 578)	Fair facing is measured separately to all items of brickwork and blockwork. In this case it is measured by doubling the area of the wall, deducting the openings and adding the area of reveals etc.
2/2/2/	5·75 3·20	147·20	(both sides of gross brickwk measurement)	
2/2/	6·15 3·20	78·72		
2/	0·20 2·20	0·88	(personnel door) (reveals	
2/	0·20 2·55	1·02	(main door)	
2/3/	0·20 1·20	1·44	(windows)	
2/	0·20 0·75	0·30		
2/5/	0·20 0·60	1·20	(opengs)	
	0·20 0·60	0·12		
2/2/2/	0·10 3·20	2·56	(pilasters)	
2/2/	0·10 2·55	1·02		
2/	0·20 2·65	1·06	(sides of wall thickening under padstone)	
		235·52		

Pumping Station 1

Timesing	Dimensions	Squaring	Description	Explanatory Notes
			Surface features (cont)	
			Ddt	
			Ditto (SW Elev	
2/	1·70 0·60	2·04	(opngs)	
2/	2·40 0·60	2·88		
2/	1·20 2·65	6·36	(attached pilasters)	
2/	0·75 1·20	1·80	(window) (NE Elev	
2/	1·80 2·55	9·18	(door)	
2/	2·20 0·40	1·76	(lintels to same)	
2/	1·10 0·20	0·44		
2/	1·00 2·20	4·40	door (SE Elev	
2/2/	0·75 1·20	3·60	(windows)	
2/	0·60 0·75	0·90		
2/	1·40 1·20	3·36	(lintels to same) (NW Elev	
2/2/	1·10 0·20	0·88		
2/4/	2·40 0·60	11·52	(opngs)	
		49·12		

Pumping Station 1

Timesing	Dimension	Squared	Ancillaries	Explanatory Notes
			Add	
			Dampproof course; pitch polymer; 200 wide (U582)	It is necessary to state the nature and width of damp proof courses (A8). Consequently the length of damp proof course over the wall thickening and pilaster has to be classified separately. The damp proof course does not carry through the door openings and is deducted.
2/2/	5.75	23.00		
2/	6.15	12.30		
Ddt		35.30		
1.20	1.20		(for door opngs)	
1.80	1.80			
6/ 0.40	2.40		(for pilasters)	
1.20	1.20	6.60	(for wall thickening to SW elevation)	
		28.70		
			Ditto 300 wide (pilasters) (U582.1)	
6/	0.40			
		2.40		
			Ditto 400 wide (U582.2) (wall thickening to SW elevation)	
	1.20			
		1.20		

Pumping Station 1

		Ancillaries (cont)	Explanatory Notes
		Building in pipes and ducts X-Sect area 0·16m² (U588)	
1		(vent fan duct.	
	1		
		Painting	
		1 coat calcium plumbate primer 2 cts gloss paint to metal sections (V1$470)	The girth calculated for the crane rail will be slightly overmeasured as it does not deduct the area where the web meets the flange or allow for the curves etc. However, it is considered to be sufficiently accurate for measurement purposes
11·80 1·52		(crane rail) girth: 4/178 712; 2/406 812; 1524	
	17·94		

Chapter 4
PUMPING STATION NO. 2

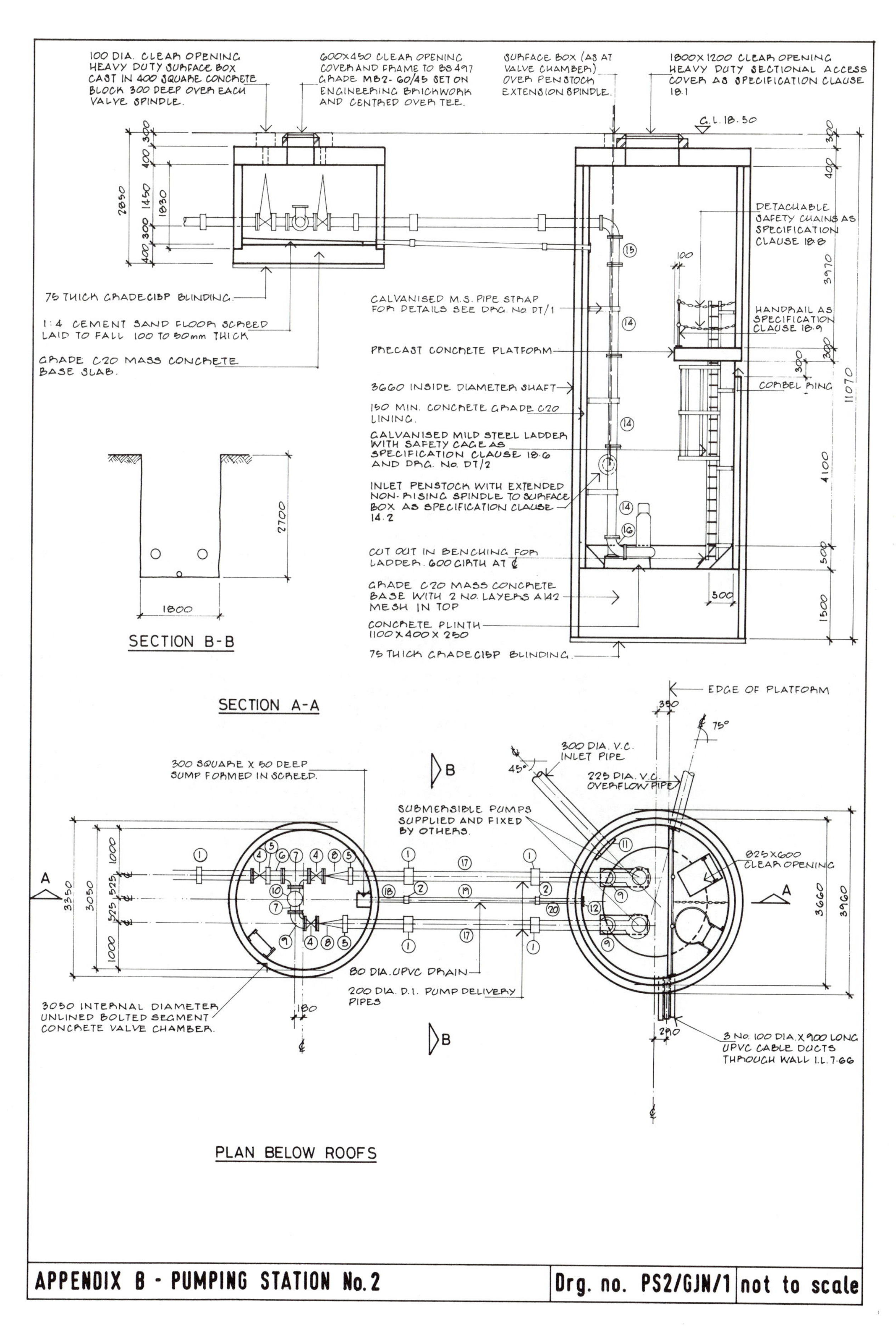
100 DIA. CLEAR OPENING HEAVY DUTY SURFACE BOX CAST IN 400 SQUARE CONCRETE BLOCK 300 DEEP OVER EACH VALVE SPINDLE.
600X450 CLEAR OPENING COVER AND FRAME TO BS 497 GRADE MB2-60/45 SET ON ENGINEERING BRICKWORK AND CENTRED OVER TEE.
SURFACE BOX (AS AT VALVE CHAMBER) OVER PENSTOCK EXTENSION SPINDLE.
1800X1200 CLEAR OPENING HEAVY DUTY SECTIONAL ACCESS COVER AS SPECIFICATION CLAUSE 18.1
G.L. 18.50
75 THICK GRADE C15P BLINDING.
1:4 CEMENT SAND FLOOR SCREED LAID TO FALL 100 TO 50mm THICK
GRADE C20 MASS CONCRETE BASE SLAB.
GALVANISED M.S. PIPE STRAP FOR DETAILS SEE DRG. No. DT/1
PRECAST CONCRETE PLATFORM
3660 INSIDE DIAMETER SHAFT
150 MIN. CONCRETE GRADE C20 LINING.
GALVANISED MILD STEEL LADDER WITH SAFETY CAGE AS SPECIFICATION CLAUSE 18.6 AND DRG. No. DT/2
INLET PENSTOCK WITH EXTENDED NON-RISING SPINDLE TO SURFACE BOX AS SPECIFICATION CLAUSE 14.2
CUT OUT IN BENCHING FOR LADDER. 600 GIRTH AT ℄
GRADE C20 MASS CONCRETE BASE WITH 2 No. LAYERS A142 MESH IN TOP
CONCRETE PLINTH 1100 X 400 X 250
75 THICK GRADE C15P BLINDING.
DETACHABLE SAFETY CHAINS AS SPECIFICATION CLAUSE 18.8
HANDRAIL AS SPECIFICATION CLAUSE 18.9
CORBEL RING
SECTION B-B
SECTION A-A
EDGE OF PLATFORM
300 SQUARE X 50 DEEP SUMP FORMED IN SCREED.
300 DIA. V.C. INLET PIPE
225 DIA. V.C. OVERFLOW PIPE
SUBMERSIBLE PUMPS SUPPLIED AND FIXED BY OTHERS.
825X600 CLEAR OPENING
80 DIA. UPVC DRAIN
200 DIA. D.I. PUMP DELIVERY PIPES
3050 INTERNAL DIAMETER UNLINED BOLTED SEGMENT CONCRETE VALVE CHAMBER.
3 No. 100 DIA. X 900 LONG UPVC CABLE DUCTS THROUGH WALL I.L. 7.66
PLAN BELOW ROOFS
APPENDIX B - PUMPING STATION No. 2
Drg. no. PS2/GJN/1
not to scale

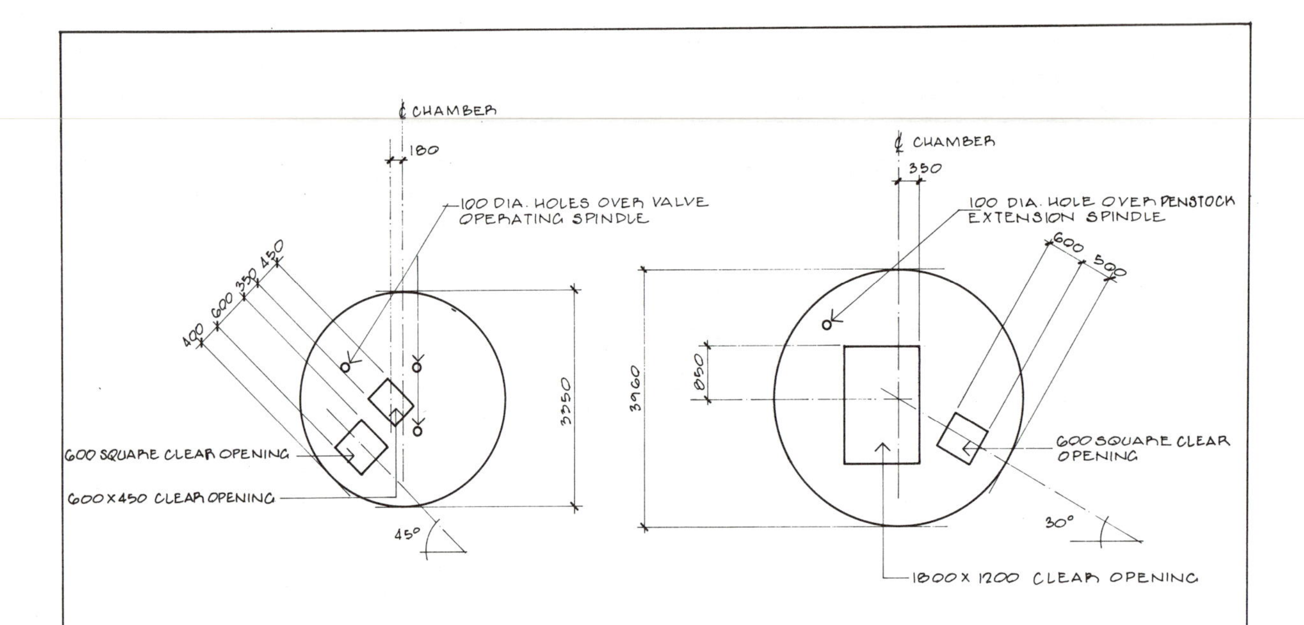

PLAN ON ROOF SLABS

ITEM No.	No. OFF	NOM. DIA.	DESCRIPTION
1	5	200	DETACHABLE FLEXIBLE COUPLING
2	2	80	DETACHABLE FLEXIBLE COUPLING
3	1	200	FLANGE SPIGOT 1200 LONG
4	3	200	DOUBLE FLANGE RISING SPINDLE SLUICE VALVE WITH GEARBOX AND EXTENSION SPINDLE TO GIVE NON-RISING VALVE OPERATING CAP APPROX. 100 BELOW SOFFIT OF ROOF SLAB.
5	3	200	FLANGE ADAPTOR
6	1	200	FLANGE SPIGOT 250 LONG.
7	2	200	ALL FLANGE EQUAL TEE
8	2	200	DOUBLE FLANGE CHECK VALVE
9	3	200	DOUBLE FLANGE 90° BEND
10	1	200	BLANK FLANGE
11	1	300	PENSTOCK AND EXTENSION SPINDLE SUITABLE FOR AN OFF SEATING PRESSURE OF 8.5M OF WATER FOR BOLTING TO THE STRUCTURE
12	1	80	FLAP VALVE FOR BOLTING TO THE STRUCTURE
13	2	200	FLANGE SPIGOT 1300 LONG
14	6	200	DOUBLE FLANGE STRAIGHT 2000 LONG
15	2	200	DOUBLE FLANGE STRAIGHT 605 LONG
16	2	200	DOUBLE FLANGE DUCKFOOT BEND
17	2	200	DOUBLE SPIGOT STRAIGHT 2900 LONG
18	1	80	DOUBLE SPIGOT UPVC STRAIGHT 750 LONG
19	1	80	DOUBLE SPIGOT UPVC STRAIGHT 3000 LONG
20	1	80	DOUBLE SPIGOT UPVC STRAIGHT 1000 LONG

FITTINGS SCHEDULE

No. OFF	CLEAR OPENING	DESCRIPTION
1	1800 X 1200	HEAVY DUTY ACCESS COVER
2	600 X 600	HEAVY DUTY HINGED LIFT OUT ACCESS COVER
1	600 X 450	HEAVY DUTY COVER
4	100 DIA.	HEAVY DUTY SURFACE BOX.

COVERS SCHEDULE

NOTES

1. ALL DIMENSIONS ARE IN MILLIMETRES AND LEVELS IN METRES
2. ROCK HEAD LEVEL AT APPROXIMATELY 10.35 M
3. SHAFT TO BE PRECAST CONCRETE BOLTED SEGMENTAL LININGS
4. ALL EXPOSED CONCRETE TO HAVE A TYPE 'B' FINISH
   ALL UNEXPOSED CONCRETE TO HAVE A TYPE 'A' FINISH

APPENDIX B - PUMPING STATION No. 2	Drg. no. PS2/GJN/2	not to scale

Pumping Station 2

Explanatory Notes

Drawing Numbers PS2/GJN/1
PS2/GJN/2

The following worked example shows a second type of pumping station constructed from bolted precast concrete shaft rings.

It is assumed that the existing site is level with no topsoil. Rockhead is at level 10.350 and the existing site levels are to be maintained.

This example demonstrates measurement in accordance with Classes E, F, G, H, J, N and T. As all the structure is below ground it is not necessary to divide it into substructure and superstructure. The work measured includes only for pipework which runs between and in the shafts. Inlet and overflow pipework are not included.

Pumping Station 2

			TUNNELS	Explanatory Notes
			Exc. straight shaft in mat. other than topsoil, rock or artificial hard mat.; dia 3·96m (T143)	As there are no payment lines shown on the drawings the excavation is measured to the net dimensions of the volumes to be excavated (M2). Due to the small size of the work, it is considered that the classification for 'other stated material' should be kept as simple as possible and the description of the material is taken from Class E. Detailed annotation of the dimensions is not necessary as it is quite apparent from the drawings where they apply.
π/	1·98 1·98 8·15		dep g.l. 18500 rock head 10350 8150	
		100·39		
			Exc. straight shaft in rock; dia. 3·96m (T133)	
π/	1·98 1·98 3·00		dep total dep 11070 blinding 75 11145 Less rock 8150 2995	
		36·95		
			Exc. straight shafts in mat. other than topsoil, rock or artificial hard mat.; dia. 3·35m (T143·1)	Although the smaller of the two shafts falls in the same third division classification as the larger shaft, it is necessary to keep it separate because the actual external diameter must be stated.
π/	1·68 1·68 2·93		dep 2850 blinding 76 2926	
		25·98		

Pumping Station 2

Timesing	Dimension	Squaring	Description	Explanatory Notes
π/	1.98 1.98 0.30	3.70	Exc. straight shaft in mat. other than topsoil, rock, or artifical hard mat.; dia. 3.96m used as filling above cover slab. (T143.2)	This is the approximate volume of material required for filling above the shaft cover slabs. A4 requires item descriptions for excavated material used as filling to be so stated and consequently it has to be measured separately.
Ddt 1.80 1.80 0.30	0.97			
0.60 0.60 0.30	0.11		(ddts for openings)	
π/ 0.05 0.05 0.30	0.01	1.09	&	
		2.61	Ddt Exc. straight shaft in mat. other than topsoil, rock or artificial hard mat.; dia 3.96m a.b.d (T143)	The volume measured for filling has to be deducted from the overall volume for excavation. The balance of material is deemed to be disposed of off site. (C1)

Pumping Station 2

Explanatory Notes

Timesing	Dimension	Squaring	Description
			Add Exc. straight shaft in mat. other than topsoil, rock or artificial hard mat.; dia. 3·35m used as filling above cover slab. (T143·3)
π/	1·68 1·68 0·30	2·66	
Ddt 0·60 0·60 0·30	0·11		(ddts for opngs)
0·60 0·45 0·30	0·08		&
3/π/ 0·05 0·05 0·30	0·01	0·20	
		2·46	Ddt Exc. straight shaft in mat. other than topsoil, rock or artificial hard mat.; dia 3·35m a.b.d. (T143.1)

Timesing	Dimensions	Squaring	Description
			Exc. surfs. in mat. other than topsoil, rock or artificial hard mat.; filling voids with grout. C.S. Spec. Clause 24.10 (T180)
π/	3.96 8.15	101.40	(large shaft sides)
π/	3.35 2.93	30.84	(small shaft sides)
		132.24	
			Exc. surfs. in rock; filling voids with grout. C.S. Spec. Clause 24.10 (T170)
π/	3.96 3.00		(large shaft sides)
		37.33	
			Exc. surfs. in mat. other than topsoil, rock or artificial hardmat. to receive blinding conc. (T180.1)
π/	1.68 1.68		
		8.87	
			Exc. surfs in rock to receive blinding conc. (T170.1)
π/	1.96 1.96		
		12.07	

Explanatory Notes

It is not necessary to classify the measurements of excavated surfaces according to the external diameter and therefore the quantities for both shafts can be combined. It is necessary to state the nature of the material to be used as filling for voids (A5) and this is best done by reference to the relevant specification clause.

It is important that the base of the shaft is not forgotten, but as this will not be treated in the same way as the sides, then it must be measured separately.

Pumping Station 2

				Explanatory notes
			Preformed segmental linings to shafts	
			Precast conc. bolted rings; flanged; nom. ring width 610mm max. piece weight 0·27t.; 4No. O segments, 2No. T segments, 1No key segment; 34No. 7" x 7/8" circle and cross bolts; 2No 10½" x 7/8" key bolts; 3mm bituminous packing to longitudinal joints; intl. dia. 3·60m	All the details of the constituent components of the shaft rings would be contained in the relevant manufacturers catalogue or in the specification. The number of rings is calculated by dividing the total shaft depth by the nominal ring width. One of the rings has a corbel and this is measured separately. The diameter given is the internal diameter.
			(TG13)	
16/	1		No 1500 500 4100 300 3970 610)10370 17No corbel ring 1No 16No.	
		16		
			Ditto but with corbel ring for half the dia.	
	1			
		1		

				Explanatory Notes
			Preformed segmental linings (cont)	
			Precast conc. bolted rings; flanged; nom. ring width 610mm; max. piece weight 0·25t; 4 No O segments, 2 No T segments, 1 No key segment; 28 No 6½" x ¾" circle and cross bolts; 2 No 10½" x ¾" key bolts; 3mm bituminous packing to longitudinal joints; intl. dia. 3.05m	
			(T613.1)	
	3/1		No 610) 1830 3 No	
		3		

Pumping Station 2

	Dimensions	Quantities	Description	Explanatory Notes
			Lining ancillaries	
			Caulking with mat. to Spec. Clause 24.8 (T674)	There is no requirement to distinguish between longitudinal and circumferential caulking and they are therefore measured together providing the materials are the same, and they are the same size. It is not a requirement to state the size, but this should not be automatically discounted, particularly if there are various sized caulking grooves.
6/	10.37	62.22	(longitudinal caulking)	
6/	1.83	10.98		
π/16/	3.66	184.00	(circumferential caulking)	
π/2/	3.05	19.17		
		276.37		
			Support and stabilization	
			Pressure grouting	No permanent support is required. It is considered undesirable to measure temporary support and M8 would have to be suitably amended in the Preamble. It is further considered that the number of sets of drilling and grouting plant should be left up to the Contractor's discretion and therefore item T831 would have to be taken out. The Specification requires the grouting to be done in two stages. The first stage is to drill to a depth of 4m and then grout. The second stage is to drill a further 2m and grout.

Pumping Station 2

				Explanatory Notes
			Support and Stabilization (cont)	
			Pressure grouting (cont)	
			Face packers (T832)	The specification calls for 3 injection points per ring. As each injection point is injected twice, the number of face packers measured is double the number of injection points (M10)
2/17/3/	1	102	(large shaft)	
2/3/3/	1	18	(small shaft)	
		120		
			Drilling and flushing 100mm dia. in holes of dep. n.e. 5m (T834)	The length of the holes has to be stated in stages of 5m (A17). As the initial stage of drilling is 4m this is classified as not exceeding 5m.
17/3/	4.00	204.00	(large shaft)	
3/3/	4.00	36.00	(small shaft)	
		240.00		
			Re-drilling and flushing 100mm dia. in holes of dep. 5-10m (T835)	For the second stage of grouting, the total hole depth is 6m, ie. stage 5-10m, and this is used for the classification of the re-drilling item, even though the re-drilling depth is only 4m. The depth of the re-drilling is the same as the initial drilling, with the second 2m stage being classed as 'initial' drilling in depth of hole 5-10m
17/3/	4.00	204.00	(large shaft)	
3/3/	4.00	36.00	(small shaft)	
		240.00		

Pumping Station 2

Timesing	Dimension	Squaring	Description	Explanatory Notes
			Support and stabilization (cont)	
			Pressure grouting (cont)	
			Drilling and flushing 100mm dia. in holes of depth 5-10m (T834.1)	
17/3/	2.00	102.00	(large shaft)	
3/3/	2.00	18.00	(small shaft)	
		120.00		
			Injection of grout to Spec. Clause 24.9 No of rings large shaft 17 small shaft 3 20 x 1t/ring 20t	The total tonnage of grout measured will be an estimate of the likely quantity based on the Engineer's knowledge of the prevailing ground conditions. The quantity inserted here is based on 1 tonne of dry material per ring.
	20t			
		20t		

ALL THE FOLLOWING WORK IN SHAFTS

				Explanatory Notes
			IN-SITU CONCRETE	Due to the fact that there are several different classifications of concrete involved, it is considered that it is more appropriate to measure the in-situ concrete and ancillaries in accordance with Classes F and G.
			Provn of conc.	
			Standard mix, ST3, OPC to B.S. 146; 20agg. to B.S. 882	
			(F133)	Because the concrete is measured in Class F instead of Class T it would be prudent to state in a general heading that all the work is in shafts
	1·70			
		1·70m^3	(blinding)	
			Designed mix grade C20; OPC to B.S.146; 20agg to B.S. 882.	
	18·48			
	3·43		(F243)	
	30·60		(structure)	
Ddt 10·77				
	1·04			
		42·78m^3		

Pumping Station 2

Timesing	Dimensions	Squaring	Description	Explanatory Notes
			IN-SITU CONCRETE (CONT)	
			Placg. of mass conc.	
			Blinding; thness n.e. 150mm (F511)	
π/	1.98			
	1.98			
	0.08	0.99	(large shaft)	
π/	1.68			
	1.68		(small shaft)	
	0.08	0.71		
		1.70		
			Bases, footgs, pile caps & ground slabs thness ex. 500mm (F524)	Although the base to the large shaft has two layers of mesh in the top it is not considered 'reinforced' for the purposes of the concrete classification.
π/	1.98			
	1.98		(bottom of large shaft)	
	1.50			
		18.48		
			ditto 300-500mm (F523)	A deduction is made from the volume where the bottom ring 'sits' in the top of the base because the cross sectional area exceeds $0.01 m^2$ (M1(d))
π/	1.68		(bottom of small shaft)	
	1.68			
	0.40	3.55		
Ddt π/ 3.20				
0.15				
0.08	0.12			
		3.43		

Side calculations:

dep
2850
300 / 400 — 700
2150
1830
320

400
320
80

dia
3350
3050
2) 6400
3200

Timesing	Dimensions	Squaring	Description
			IN-SITU CONCRETE (CONT)
			Placg. of mass conc. (cont)
			Lining to shaft walls; intl dia. 3360 mm; min. thness 150mm (F580)
π/	3.66		
	8.87		
	0.30	30.60	
			Ddt ditto
14.54/	0.71		
	1.00		
	1.00	10.32	(rings)
	5.05		(corbel & platform)
	0.15		
	0.60	0.45	
		10.77	

Waste calculations:

Len
500
4100
300
3970
8870

Dia.
extl. 3960
intl 3360
2) 7320
3660

Wid
rings 150
lining 150
300

No. rings
610) 8870
14.54

Vol.
$0.71 m^3$

Len
½/π/3.660 5749
2/350 700
5049.

Explanatory Notes

The minimum thickness of the lining is 150mm. This thickness is deducted from the internal diameter of the shafts to give the internal diameter of the lining

The volume of the lining is derived by calculating the total volume of the lining including the space occupied by the rings, and then deducting the volume of the rings which is stated in the manufacturer's literature. This is the easiest way of taking into account the flanges in the rings. Further adjustments are made for the volume occupied by the corbel and the platform.

Pumping Station 2

				Explanatory Notes
			IN-SITU CONCRETE (CONT)	
			Placg. of mass conc. (cont)	
			Benching; 500 x 500 o/all; bott. of shaft (F580.1)	The volume of the benching is calculated by multiplying its cross-sectional area by the mean diameter. The volume of the cut out for the ladder is deducted, but the overlap with the concrete plinths is not.
1/2/π/	2.86		dia 3360	
	0.50		2/1/2/500 500	
	0.50	1.12	2860	
Ddt 1/2/ 0.60				
0.50			(for cut out for ladder)	
0.50	0.08			
		1.04		

Timesing	Dimension	Squared	Description	Explanatory Notes
			CONCRETE ANCILLARIES	
			Fmwk. type A fin	
			Plane curved to 3·35m rad. in one plane 0·2-0·4m (G153)	The radius of the curved formwork must be stated in item descriptions (A4).
	π/3·35 0·32	3·37	(extl face of base to small shaft)	As the formwork to the outside face of the base to the small shaft is exposed, it has a type A finish.
			Fmwk. type B fin.	
			Plane curved to 3·36m rad. in one plane ex 1·22m (G255)	Note that the measurements of the lining formwork does not include the area behind the benching, but does include for the area to the ladder cut out.
	π/3·36 8·37	88·36	(lining to large shaft) Len 8870	
	0·60 0·50	0·30	(ladder benching cut out) 500 / 8370	
Ddt 4·58 0·60	2·75		(for corbel & platform) Len ½/π/3360 5278; 2/350 700; 4578	
		85·91		
			Plane curved to varying radii; min 2360; max 3360 (G260)	It is not clear whether curved formwork which is also the upper surfaces of concrete should be so described. It is considered that it is not necessary in this particular case, although it should be made clear in the Preamble.
	π/2·86 0·71	6·38	(benching)	
Ddt 0·60 0·70	0·42		(for ladder cut out) wid $x^2 = 500^2 + 500^2$; $x = 707$	
		5·96		

				Explanatory Notes
			CONCRETE ANCILLARIES (CONT)	
			Fmwk. type B fin (cont)	
			Plane. vert. average width 0·25m (G243)	This is not strictly the correct method of measuring this item, but the quantity is so small that it is considered to be more applicable than dividing the item into the various width classifications which could give inordinately high quantities.
2/½/	0·50		(sides of cut out to benching)	
	0·50	0·25		
			Reinforcement	
			Fabric to B.S. 4483 ref. A142. (G562)	The area of fabric is the net area with no allowance for fabric in laps (M9).
2/π/	1·98		(base to large shaft)	
	1·98	24·64		

Pumping Station 2

Timesing	Dimensions	Squaring	Description	Explanatory Notes
			CONCRETE ANCILLARIES (CONT)	
			Concrete Accessories	
			Fin. of top surfs; steel trowel fin.	The specification calls for a steel trowel finish to the base of the large shaft. No deduction is made for the support plinths as their plan area does not exceed 0·5 m² (M14)
π/	1·18 1·18	4·37	(base to large shaft) (G812) radius 2)2360 = 1180	
	0·60 0·50	0·30	(ladder cut outs)	
		4·67		
			C. & S. (1:4) screed; av. 75mm th. with steel trowel fin. (G815)	A13 requires the materials, thickness and surface finish of applied finishes to be stated. No deduction is made for the sump as this is less than 0·5m². The forming of the sump could be measured separately or as in this case Coverage Rule C6 could be extended to include it.
π/	1·53 1·53	7·36	radius 2)3050 = 1525	
			Inserts	
			80mm dia. pipe through precast conc. shaft linings; proj. both sides (supply incl. elsewhere) (G832)	
	1	1	(drain i small shaft)	

Pumping Station 2

				Explanatory Notes
			CONCRETE ANCILLARIES (CONT)	
			Inserts (cont)	
			200mm dia. pipe through precast conc. shaft linings; proj. both sides (supply incl. elsewhere) (G832.1)	One area not covered by Class T is the work involved in breaking through linings for incoming pipes and similar inserts. It is therefore necessary to create new items which inform the Contractor of the work involved.
	3/ 1	3	(pump pipes - small shaft)	
			All the following inserts projecting both sides of precast conc. shaft segments with in-situ conc. lining	
			100mm dia x 900mm long U.P.V.C cable ducts (G 832.2)	The supply and fixing of the ducts is included in this item (C7)
	3/ 1	3	(large shaft)	
			80mm dia. pipe; (supply included elsewhere) (G832.3)	
	1	1	(drain - large shaft)	

Pumping Station 2

				Explanatory Notes
			CONCRETE ANCILLARIES (CONT)	
			All the following inserts projecting (cont)	
			200 mm. dia. pipe (supply incl. elsewhere)	
	2/ 1		(G832·4) (pump delivery pipes - large shaft)	
		2		
			225 mm dia. pipe (supply incl. elsewhere)	
	1		(G832·5) (overflow pipe - large shaft)	
		1		
			300 mm dia. pipe (supply incl. elsewhere)	
	1		(G832·6) (inlet pipe - large shaft)	
		1		

Pumping Station 2

				Explanatory Notes
			PRECAST CONCRETE	
			Slabs; conc. designed mix grade C30, OPC to B.S. 146; 10 agg to BS. 882.	A1 requires the Specification of concrete to be stated in items for precast concrete. The Second and Third divisions of the slab's classification require the area and weight of the slabs to be stated in bands as classified. These only have to be approximate, as the Contractor will ascertain from the drawings the exact nature of the work involved.
			Areas	
			large cover slab	
			$\pi/1.98$	
			1.98	
			$12.31m^2$	
			small cover slab	
			$\pi/1.68$	
			1.68	
			$8.86m^2$	
			platform	
			$1/2/\pi/1.83$	
			1.83 5.26	
			less	
			0.35	
			1.83 0.64	
			$4.62m^2$	
			Weights	
			12.31	
			0.40	
			$2.40t/m^3$	
			11.82t	
			8.86	
			0.40	
			$2.40t/m^3$	
			8.51t	
			4.62	
			0.30	
			$2.40t/m^3$	
			3.33t.	

Pumping Station 2

				Explanatory Notes
			PRECAST CONCRETE (CONT)	
			Slabs (cont)	
	1	1	Roof slab to large shaft; area 4-15m²; weight 10-20t; thness 400mm (H537)	A1 requires the position in the works of each unit to be stated. The units do not have mark or type numbers and these cannot be stated in accordance with A2.
	1	1	Roof slab to small shaft; area 4-15m²; weight 5-10t; thness 400mm (H536)	
	1	1	Intermediate platform to large shaft; area 4-15m²; weight 2-5t; thness 300mm (H535)	

Pumping Station 2

Explanatory Notes

It was recommended in the text that the measurement of pipelines is considered on an individual basis for each contract. The majority of the pipes and fittings in this example are in the shafts but there are the lengths between the shafts to take into account. To try and measure these pipes as 'in trenches' would cause undue complication because the fittings into the chamber would be then part 'extra over' the pipe in the trench and part 'full value' for the length in the chamber. In this instance the recommended procedure is to measure the excavation for all three pipes separately and to enumerate all the pipes and fittings regardless of length as not in trenches. It would be necessary to state the measurement philosophy adopted in the Preamble.

Pumping Station 2

			Explanatory Notes
		EARTHWORKS TO PIPEWORK	
		Pipe tr. exc. as drwg PS2/GSN/1 mat. other than topsoil, rock or artificial hard mat.; max dep. 2-5m (E425)	There is no classification for pipe trench excavation and backfilling so these are rogue items. This is a slight overmeasurement as the length is taken as the length of the UPVC drain which lies between the inside faces of the chamber.
4·75 1·80 2·70	23·09	750 3000 1000 4750	
		&	
		Disposal; backfilling around pipes (E532)	

Times	Dimensions	Squaring	Description
			PIPEWORK - FITTINGS AND VALVES; NOT IN TRENCHES
			Ductile iron pipe fittings; B.S. 4772; flanged joints; nom. bore 200mm; as Spec. Clause 13.1
			Flanged spigot pipe 1200 mm long (J381)
	1	1	(ref 3)
			Flanged spigot pipe 250 mm long (J381.1)
	1	1	(ref 6)
			Flanged spigot pipe 1300 mm long (J381.2)
2/	1	2	(ref 13)
			Double flanged pipe 2000mm long. (J381.3)
6/	1	6	(ref 14)

Explanatory Notes

Whilst all the fittings are scheduled, the prudent taker off would still check the fittings off the drawing as they were measured.

Although the fittings are described as having flanged joints, the first fitting has a spigot one end and this tells the estimator that this end does not have a flanged joint. This end has a flexible coupling which is measured separately and the cost of the joint to the spigot end is therefore included elsewhere.

Pumping Station 2

Timesing	Dimension	Squaring	Description	Explanatory Notes
			PIPEWORK (CONT)	
			Ductile iron pipe fittings (cont)	
2/	1	2	Double flanged pipe 605mm long (J381.4) (ref 15)	
2/	1	2	Double spigot pipe 2900mm long (J381.5) (ref 17)	Although this fitting does not have flanged joints at either end it is still classed as such for convenience. The estimator will know that the flexible couplings are measured separately and that he does not have to allow for joints for this fitting.
3/	1	3	Double flanged 90° bend (J311) (ref 9)	As both these bends are less than 300mm it is not necessary to describe them as vertical.
2/	1	2	Double flanged 90° duckfoot bend (J311.1) (ref 16)	

Pumping Station 2

Timesing	Dimension	Squaring	Description
			PIPEWORK (CONT)
			Ductile iron pipe fittings (cont')
			All flanged equal tee (J321)
2/	1		(ref 7)
		2	
			Blank flange (J391)
	1		(ref10)
		1	
			Flexible coupling (J391.1)
5/	1		(ref 1)
		5	
			Flange adaptor (J351)
3/	1		(ref5)
		3	

Explanatory Notes

Pumping Station 2

Timesing	Dimension	Squaring	Description	Explanatory Notes
			PIPEWORK (CONT)	
			U.P.V.C pipe fittings; B.S. 4660; nom. bore 80mm	As the flexible couplings are measured separately it is not necessary to state the nature of the joints in the item descriptions
	1	1	double spigot pipe 750mm long. (ref 18) (J481)	
	1	1	double spigot pipe 3000mm long (ref 19) (J481·1)	
	1	1	double spigot pipe 1000mm long (ref 20) (J481·2)	
2/	1	2	Flexible coupling (ref 2) (J491·2)	

Pumping Station 2.

Timesing	Dimension	Squaring	Description	Explanatory Notes
			PIPEWORK (CONT)	
			Gunmetal valves and penstocks: as Spec. Clause 13.6	
3/	1	3	Double flanged rising spindle sluice valve with gearbox and extension spindle; nom. bore 200mm (ref 4) (J811)	The specification would have complete details of the valves and penstocks, and reference is made to the relevant clause: no attempt is made to describe them fully in the description.
	1	1	Flap valve; including all fixing to structure nom. bore 80mm (ref 12) (J891)	
2/	1	2	Double flanged. check valve, nom bore 200mm (ref 8) (J891.1)	
	1	1	Penstock with extended non-rising spindle including all brackets and fixings as Spec. Clause 14.2 (ref 11) (J881)	

Pumping Station 2.

				Explanatory Notes
			PIPEWORK - SUPPORTS AND PROTECTION	
			Concrete stools and thrust blocks; Concrete grade C20, OPC to B.S. 146, 20agg to BS 882; unreinforced	A5 requires the specification of concrete to be stated and whether it is reinforced. If there were several reinforced stools within the same volume classification, it would be necessary to itemise them individually if their reinforcement requirements were different.
			Vol. 1·1 0·4 0·25 $0·11m^3$	
			Vol. $0·1 - 0·2m^3$. nom bore 200mm	
	2/ 1	2	(large shaft) (L721)	
			Other isolated pipe supports; as drwg DT/1	Although A6 states that the principal dimensions and materials should be stated in item descriptions, the reference to the drawing would in effect be giving the same information
			height n.e. 1m. nom. bore 200mm	
	2/3/ 1	6	(large shaft) (L811)	

Pumping Station 2

			Description	Explanatory Notes
			MISCELLANEOUS METALWORK	
	1	1	Ladders; galvanised mild steel as Spec. Clause 18.6 & drwg DT/2 with safety cage; ladder 5700mm long incl. all fixing. (N130)	The unit of measurement for ladders in Class N is linear metres, but this ladder has a safety cage for part of its length which would mean measuring it as two items, one with and one without the cage. It is considered that enumerating the ladder stating the length is a fairer method of measurement.
	3.20	3.20	Handrail as Spec. Clause 18.9 incl. all fixings (N140)	
2/	1	2	Safety chain as Spec. Clause 18.8; net. length 1300mm including all fixings (N190)	There is no separate classification for safety chains and this is a rogue item. They could have been measured in linear metres as an alternative to enumeration.

Pumping Station 2

			Explanatory Notes
		MISCELLANEOUS METALWORK (CONT)	
		Access covers and frames on two course Engineering brickwork	There is no separate classification for access covers and frames, so these are rogue items.
		B.S. 497 grade MB2 - 60/45 (N190)	
1		(small shaft)	
	1		
		B.S. 497 grade MB2 - 60/60 (N190.1)	
2/ 1		(small & large shafts)	
	1		
		1800 x 1200 clear opening as Spec. Clause 18.1 (N190.2)	
1		(large shaft)	
	1		

				Explanatory Notes
			MISCELLANEOUS METALWORK (CONT)	
			Heavy duty surf. box incl casting in 400 x 400 x 300mm deep concrete	The concrete surround is given in the item description, which is considered more appropriate than measuring it in detail in accordance with Classes F & G.
			100mm dia. clear opng. (J190.3)	
3/	1	3	(for valve spindles)	
	1	1		
		4		

Chapter 5
RETAINING WALL

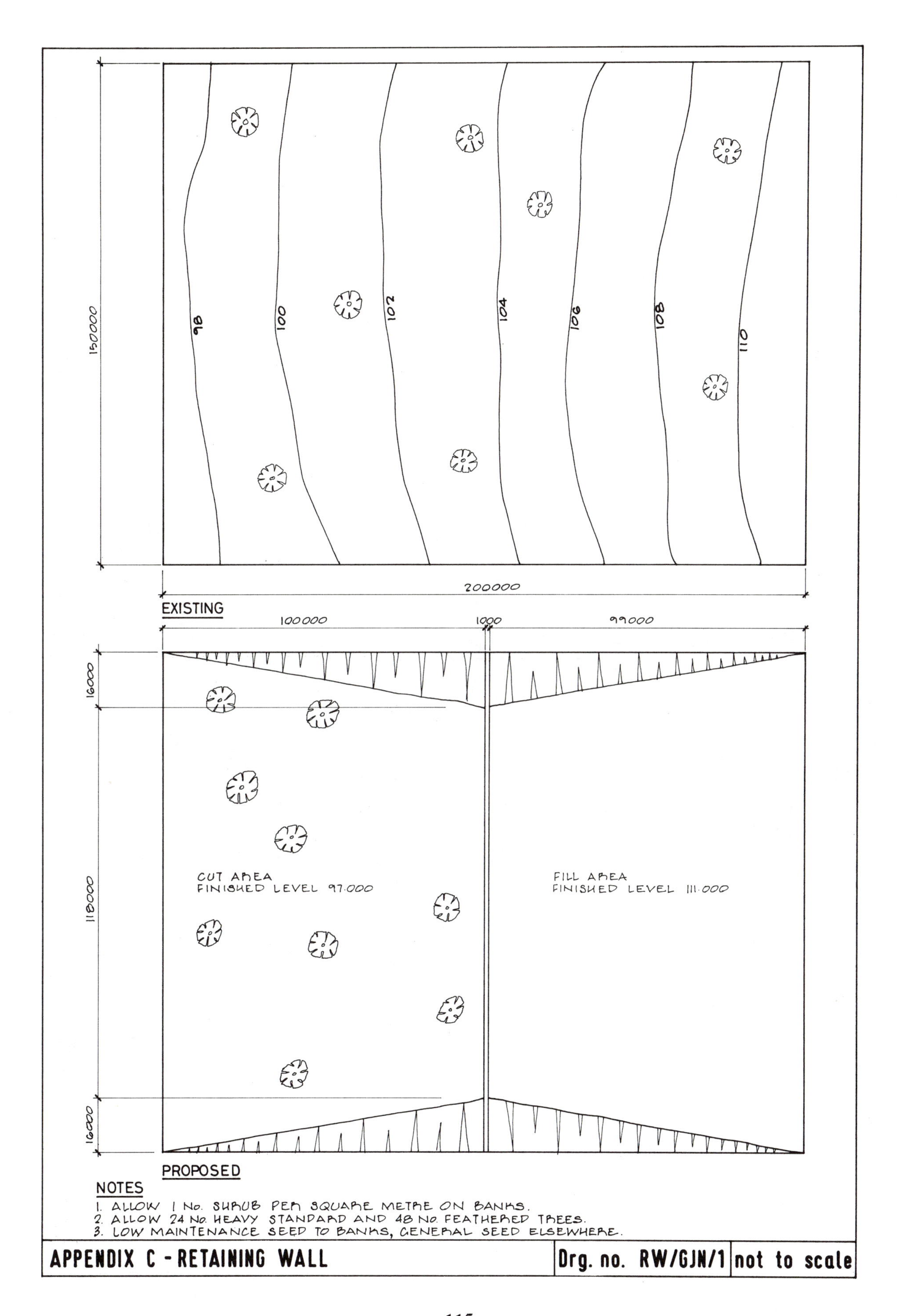
150000
98
100
102
104
106
108
110
200000
EXISTING
100000
1000
99000
16000
118000
16000
CUT AREA
FINISHED LEVEL 97.000
FILL AREA
FINISHED LEVEL 111.000
PROPOSED
NOTES
1. ALLOW 1 No. SHRUB PER SQUARE METRE ON BANKS.
2. ALLOW 24 No. HEAVY STANDARD AND 48 No. FEATHERED TREES.
3. LOW MAINTENANCE SEED TO BANKS, GENERAL SEED ELSEWHERE.
APPENDIX C - RETAINING WALL
Drg. no. RW/GJN/1
not to scale

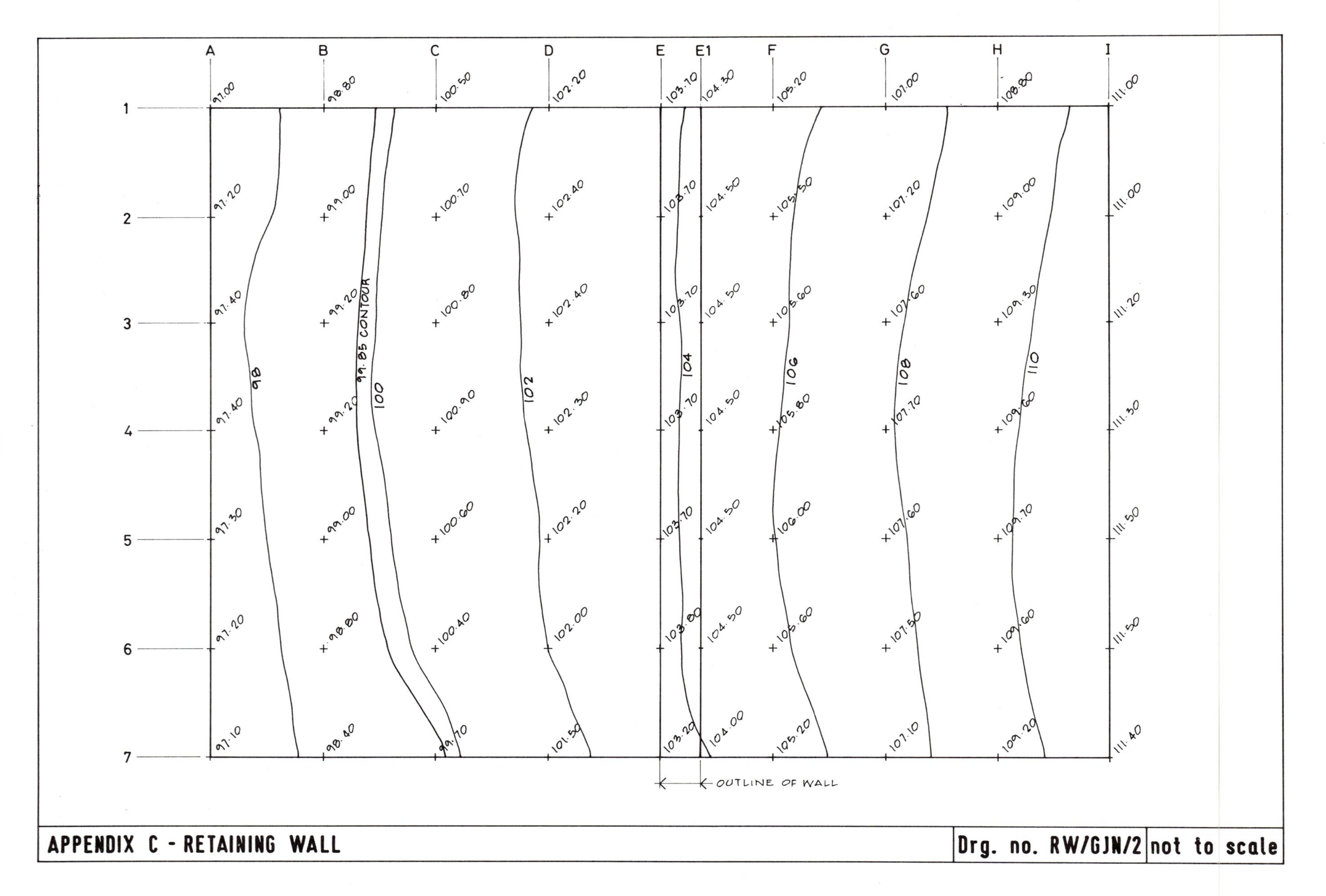

A
B
C
D
E
E1
F
G
H
I
1
2
3
4
5
6
7
97.00
98.80
100.50
102.20
103.10
104.30
105.20
107.00
108.80
111.00
97.20
99.00
100.10
102.40
103.70
104.50
105.50
107.20
109.00
111.00
97.40
99.20
100.80
102.40
103.70
104.50
105.60
107.60
109.30
111.20
97.40
99.20
100.90
102.30
103.70
104.50
105.80
107.70
109.60
111.30
97.30
99.00
100.60
102.20
103.70
104.50
106.00
107.60
109.70
111.50
97.20
98.80
100.40
102.00
103.80
104.50
105.60
107.50
109.60
111.50
97.10
98.40
99.70
101.50
103.20
104.00
105.20
107.10
109.20
111.40
98
99.85 CONTOUR
100
102
104
106
108
110
OUTLINE OF WALL
APPENDIX C - RETAINING WALL
Drg. no. RW/GJN/2
not to scale

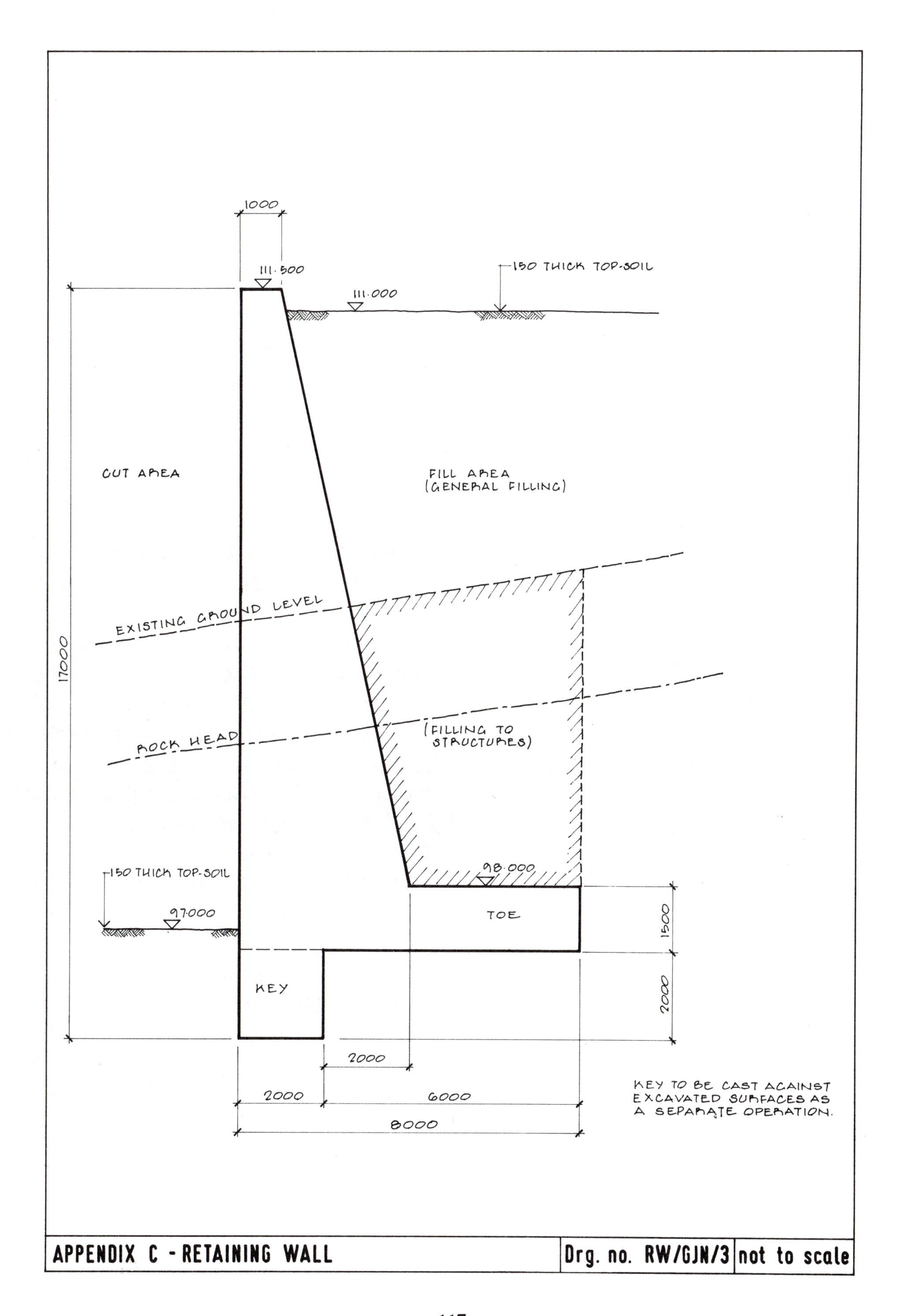
1000
111.500
111.000
150 THICK TOP-SOIL
CUT AREA
FILL AREA
(GENERAL FILLING)
EXISTING GROUND LEVEL
ROCK HEAD
(FILLING TO
STRUCTURES)
17000
150 THICK TOP-SOIL
97.000
98.000
TOE
1500
KEY
2000
2000
2000
6000
8000
KEY TO BE CAST AGAINST
EXCAVATED SURFACES AS
A SEPARATE OPERATION.
APPENDIX C - RETAINING WALL
Drg. no. RW/GJN/3
not to scale

Retaining Wall

				Explanatory Notes
	Drawing Numbers RW/GJN/1			The following example shows a rectangular virgin field with a reasonably steep gradient. The example
	RW/GJN/2			involves the construction of a major concrete retaining wall across the middle of the site to enable one half of the area to be filled to a constant level whilst lowering the other half.
	RW/GJN/3			

It is assumed that the site will have a constant 100mm of good quality topsoil together with various old trees which will have to be removed. Rock level is 3m below original ground level. This example deals only with Class E: Earthworks, the removal of the trees and the retaining wall itself would be measured under classes D, F and G.

The measurement of Earthworks invariably produces large amounts of waste calculations. If these are particular to any one item or group of items it is recommended that they are done adjacent to the item of work to which they apply as in this particular example.

Retaining Wall

Explanatory Notes

Alternatively, large amounts of general calculations may be done at the beginning of the take off with relevant annotation as to how and to which sections it applies.

For the purposes of measurement the site is divided into the following

1. Topsoil strip.
2. Excavation for the retaining wall.
3. Excavation over cut area.
4. Excavation ancillaries.
5. Filled area.
6. Topsoil and landscaping.

Retaining Wall

Explanatory Notes

1. TOPSOIL STRIP

			Description	Explanatory Notes
			gen. exc; topsoil max. dep. n.e 0·25m; exc. surf U/S topsoil (E411)	It is not necessary to state the Commencing Surface as it is also the Original Surface. The Excavated Surface and hence the maximum depth is different depending upon whether the topsoil is in the cut or filled area. However due to the nature of the work it is felt that it would be more helpful to state the Excavated Surface as underside of topsoil and therefore give the maximum depth as the depth of the topsoil in accordance with paragraph 5.21.
	200·00			
	150·00			
	0·10			
		3000·00		

The original site survey drawing RW/GJN/1 has level information by way of contours. This has been processed into a 25 metre grid of levels by interpolation as follows:-

Retaining Wall

Explanatory Notes

1. TOPSOIL STRIP (CONT)

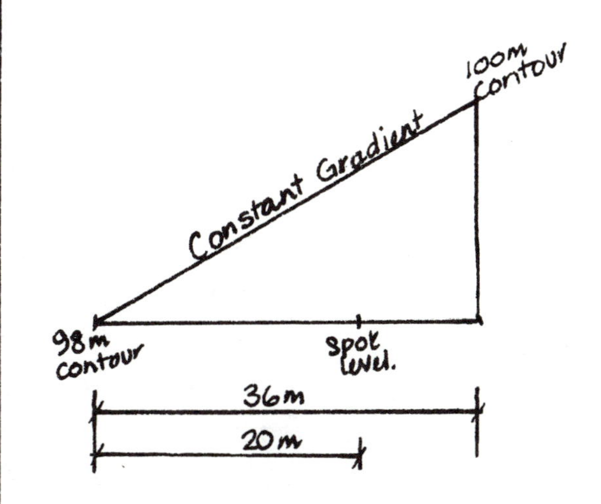

The height of the spot level is derived by calculating the vertical rise per metre of horizontal distance between the contours and multiplying it by the distance of the spot height from the lower contour. Add this value to the value of the lower contour to give the spot height e.g.

$$\text{spot height} = \left(\frac{2}{36} \times 20\right) + 98$$

$$= \underline{99.111}$$

Explanatory Notes

2. RETAINING WALL
EXCAVATION

Level		Avge G.L.
E1	1/103700	103700
E2	2/103700	207400
E3	2/103700	207400
E4	2/103700	207400
E5	2/103700	207400
E6	2/103800	207600
E7	1/103200	103200
E1.1	1/104300	104300
E1.2	2/104500	209000
E1.3	2/104500	209000
E1.4	2/104500	209000
E1.5	2/104500	209000
E1.6	2/104500	209000
E1.7	1/104000	104000
	24)	2497400
		104058
Less	topsoil	100
Average level	after topsoil strip	103958

	Form lvl.
top of toe	98000
toe thness	1500
	96500

It is an express requirement of the Specification to carry out the bottom 'key' excavation as a separate operation and concrete immediately against the excavated surface and this is deemed to be a separate stage for excavation under M5. The first part of the excavation is from the stripped level to the underside of the toe.

This is the calculation of the average ground level after topsoil strip over the retaining wall area. Drawing RW/GJN/2 has the outermost outline of the foundations shown together with the interpolated levels, (see the calculation of the average ground level for the lower side of the site for an explanation of the weighting of the levels)

Explanatory Notes

2 RETAINING WALL
EXCAVATION (CONT)

			Description	Explanatory Notes
			Max dep of excavn	
			highest spot lev. 104500	The maximum depth of the excavation must be stated and is calculated in accordance with paragraph 5.21
			topsoil 100	
			104400	
			form lev. 96500	
			7900	
			i.e. 5-10m	
			Overall av. dep of excav	
			av. lev after topsoil strip 103958	This is the calculation of the average overall depth of the excavation and will include excavation of natural material and rock. It is necessary to calculate the average depth of both in order to measure both separately.
			form lev. 96500	
			7458	
			Av. dep of natural material	
			overall av. dep. g l to top of rock 3000	
			Less topsoil 100	
			2900	The Commencing Surface in this case is the underside of the topsoil which is not also the Original Surface and therefore must be stated. The Excavated Surface is bottom of the first stage of excavation i.e. underside of the toe. As this is not also the Final Surface it also must be stated (A4). It is not necessary to state the type of material as it is deemed to be natural material other than topsoil, rock or artificial hard material (D1).
			Exc. foundns; max dep 5-10m; Comm. surf. U/S of topsoil; exc surf U/S of wall toe (lev 96500)	
	150.00		(E 326)	
	9.00			
	2.90			
		3915.00		

Retaining Wall

Explanatory Notes

2. RETAINING WALL

EXCAVATION (CONT)

Av. dep of rock

av. dep of total excav.	7458
less av. dep of nat. mat.	2900
	4558

Dimensions	Quantity	Description	Explanatory notes
150.00 9.00 4.56	6156.00	Exc. foundns; rock; max. dep. 5-10m; comm. surf. U/S of topsoil; exc. surf U/S of wall toe (lev 96500) (E 336)	Although the rock is in a different location in the excavation and of a different depth, the depth classification is exactly the same as the previous item because in accordance with definitions of 1.12 and 1.13, the Commencing and Excavated Surfaces are the same for both items, and therefore the depth in accordance with paragraph 5.21 is the same. It is the overall depth of the void which is critical, not the depth of the individual materials within it.

toe excav

Dimensions	Quantity	Description	Explanatory notes
150.00 2.00 2.00	600.00	Exc. foundns; rock; max. dep. 1-2m; comm. surf U/S of wall toe (lev. 96500) (E 334)	This is a separate stage of excavation in accordance with M5. The Commencing Surface must be stated as it is not also the Original Surface. It is not necessary to state the Excavated Surface as it is also the Final Surface (A7). The Commencing Surface may be identified by stating the level or by it's relationship to any other fixed point.

Retaining Wall

Explanatory Notes

3 CUT AREA

Level		Av. gl after topsoil strip
A1	1/97000	97000
B1	2/98800	197600
C1	2/100500	201000
D1	2/102200	204400
E1	1/103700	103700
A2	2/ 97200	194400
B2	4/ 99000	396000
C2	4/100700	402800
D2	4/102400	409600
E2	2/103700	207400
A3	2/ 97400	194800
B3	4/ 99200	396800
C3	4/100800	403200
D3	4/102400	409600
E3	2/103700	207400
A4	2/ 97400	194800
B4	4/ 99200	396800
C4	4/100900	403600
D4	4/102300	409200
E4	2/103700	207400
A5	2/ 97300	194600
B5	4/ 99000	396000
C5	4/100600	402400
D5	4/102200	408800
E5	2/103700	207400
A6	2/ 97200	194400
B6	4/ 98800	395200
C6	4/100400	401600
D6	4/102000	408000
E6	2/103800	207600
	c/f	8853500

This is the calculation of the average ground level of the lower side of the site after topsoil strip.

The grid of levels extends from A-E / 1-7. The corner levels (A1) only have on sphere or 'square' of influence and are used only once. The side levels (B1, C1) have two spheres or 'squares' of influence and are used twice and the internal levels have four spheres or 'squares' of influence and are used four times. The total is then divided by the total number of spheres or 'squares' of influence to give the average level.

A B C
1
2
3

Retaining Wall

Explanatory Notes

3. CUT AREA (CONT)

Level

		b/f 8 853 500
A7	1/97100	97100
B7	2/98400	196800
C7	2/99700	199400
D7	2/101500	203000
E7	1/103200	103200
	96)	9653000

Av. Ex. g.l.	100552
topsoil	100
	100452

Final surf lev

	Fin lev.	97000
Less	topsoil thness	150
		96850

The level to which excavation is to be taken to is the finished site level less the thickness of topsoil to be placed

Max dep of excav

highest spot lev	103800
topsoil	100
	103700
final surf lev	96850
	6850

ie. 5-10m

The maximum depth of the excavation must be stated and is calculated in accordance with paragraph 5.21

Overall av. dep. of excav.

av. gl after topsoil strip	100452
final surf lev	96850
	3602

Explanatory Notes

3. CUT AREA (CONT)

			Description	Explanatory notes
	100.00 150.00 3.60	54000.00	Gen. exc.; max dep. 5–10m; comm surf U/S of topsoil (E426)	It is not necessary to state the Excavated Surface as this is also the Final Surface. The Commencing Surface must be stated as it is not also the Original Surface (A4) This quantity is overmeasured as it also contains the rock excavation which has to be measured separately and because it also assumes that the sides of the excavation are vertical when in fact they are sloping. Unlike the excavation for the foundations to the wall itself where the natural material and rock were measured separately, here it is easier if the total excavation is measured as if it were all natural material and the rock then measured and deducted from the gross quantity of natural material.

3. CUT AREA (CONT)

Level		Av. rockhead lev.
C1	1½/100500	150750
C2	3/ 100700	302100
C3	3/ 100800	302400
C4	3/ 100900	302700
C5	3/ 100600	301800
C6	2½/ 100400	251000
C7	1/ 99700	99700
D1	2/ 102200	204400
D2	4/ 102400	409600
D3	4/ 102400	409600
D4	4/ 102300	409200
D5	4/ 102200	408800
D6	4/ 102000	408000
D7	2/ 101500	203000
E1	1/ 103700	103700
E2	2/ 103700	207400
E3	2/ 103700	207400
E4	2/ 103700	207400
E5	2/ 103700	207400
E6	2/ 103800	207600
E7	1/ 103200	103200
	53)	5407150
		102022
Less dep to rock head		3000
		99022

Explanatory Notes

As stated, the previous excavation item will include rock excavation for which adjustment will have to be made and it is therefore necessary to determine the area and depth over which the rock excavation will occur. In our example the rock head is 3 metres below original ground level. If we take a section through this area we

original surface
rock head
99850
3m
final surface
96850

need to determine the line at which the final surface level intersects the rock head. This is most easily done by adding 3m onto the final surface level (96850) and plotting this contour on the original ground levels plan by interpolation (see RW/GSN/2). The area bounded by the 99850 contour and the retaining wall is the plan area of rock.

Retaining Wall

Explanatory Notes

3 CUT AREA (CONT)

Timesing/Dimensions	Squaring	Description	Explanatory Notes
		Av. dep of rock	
		av. rock lev 99022	The average rock depth is calculated by deducting the Final Surface from the average rock level as shown opposite.
		Fin. surf 96850	
		2172	
		Gen exc; rock; max. dep. 5–10m; comm. surf u/s of topsoil (E436)	The easiest method of calculating the rock area is to divide it up into equal 25m strips as on the grid and measure the area of each.
63·00 25·00 2·17	3417·75	&	
65·00 25·00 2·17	3526·25	Ddt Gen exc; max. dep 5–10m; comm surf u/s of topsoil a.b.d. (E426)	This is the deduction of the rock volume from the total quantity of the general excavation in natural materials. a.b.d. stands for 'as before described' and shows that the item has already been measured.
67·00 25·00 2·17	3634·75		
66·00 25·00 2·17	3580·50		
62·00 25·00 2·17	3363·50		
55·00 25·00 2·17	2983·75		
	20506·50		

Retaining Wall

			Total vol of both pyramids
			2 × 1/3 × 1/2 × 16 × 6.50 × 100
			= 3467 m³
			Total vol. of rock within both pyramids
		2 × 1/2 ×	(8.90 × 3.50 × 54.00) +
		2 × 1/3 ×	(0.5 × 7.1 × 3.50 × 54.00)
			= 2130 m³
			∴ Vol. of nat. mat.
			3467.00
			2130.00
			1337.00 m³
			Ddt
			Gen exc; max. dep 5-10m; comm. surf U/S topsoil a.b.d.
	1337.00		(E426)
	1.00		
	1.00	1337.00	
			Ddt
			Gen. exc; rock; max dep. 5-10m; comm surf. U/S topsoil a.b.d.
	2130.00		(E436)
	1.00		
	1.00	2130.00	

Explanatory Notes

Adjustment must also be made to the general excavation in natural material quantities for the two sloping banks on either side which are not excavated. These are in the rough shape of a triangle based pyramid as follows:-

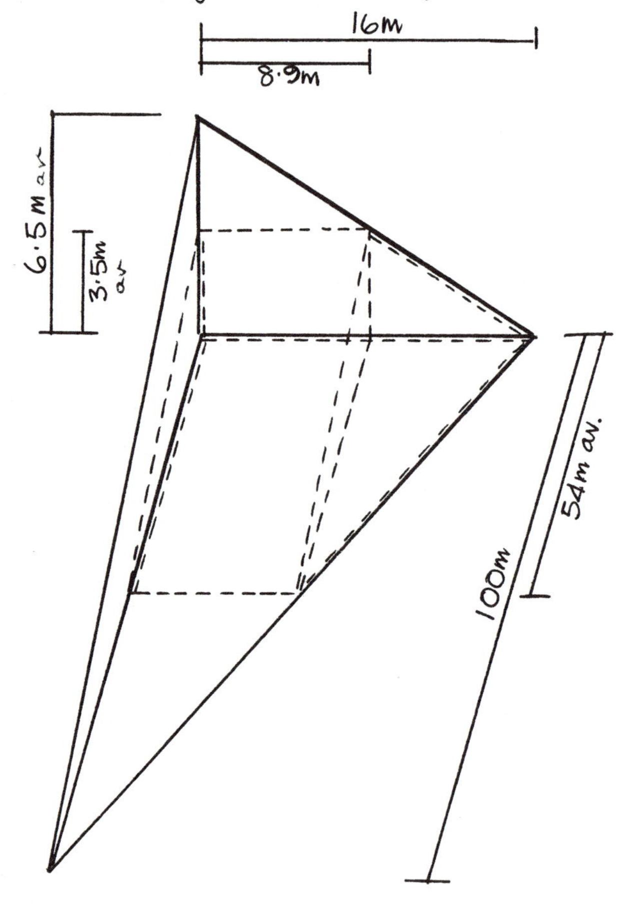

With the volume of rock being enclosed by the dotted lines.

Retaining Wall

Explanatory Notes

4. EXCAVATION ANCILLARIES

Timesing	Dimensions	Squaring	Description	Explanatory Notes
			Add	
	150.00 9.00	1350.00	Prep. of surfs; rock (E 523)	Bottom of retaining wall excavation. No preparation is measurable over the general cut area as it is to receive topsoil (M11)
2/	150.00 2.00		Prep. of surfs; rock; vert. (E 523.1)	Either side of key where the concrete is cast against excavated surfaces.
2/	9.00 2.00	636.00	(ends)	
	150.00 9.00 4.56	6156.00	Disposal of exc. mat; rock (E 533) (wall fndnc)	In our example, the rock material is deemed not to be suitable for the general filling behind the retaining wall and therefore removed from site
	150.00 2.00 2.00	600.00		
	63.00 25.00 2.17	3417.75		
	65.00 25.00 2.17	3526.25		
	67.00 25.00 2.17	3634.75	(gen exc.)	
	66.00 25.00 2.17	3580.50		
	62.00 25.00 2.17	3363.50		
	55.00 25.00 2.17	2983.75		
		27262.51		

Retaining Wall

Explanatory Notes

4. EXCAVATION ANCILLARIES (CONT)

Timesing	Dimensions	Squaring	Description	Explanatory Notes
			Ddt	
			Disposal of exc. mat; rock	
	2130.00 1.00 1.00		(E533)	Adjustment for the banks.
		2130.00		
				There are no trimming or preparation items measurable over the general cut area as all these surfaces are to receive topsoil (M10 & M11)

Explanatory Notes

5. FILLED AREA

Mat. available

ddt	Add
	3915.00
	54000.00
20506.50	
1337.00	
21843.50	57915.00
	21843.50
	36071.50 m^3

There are two types of filling involved. The first is backfilling to the structure above the toe level to bring the levels up to the stripped level, and then general filling to the required finished levels. The first calculation should be to compute the amount of excavated materials available for filling.

Fill. to structure

dep.

av. lev. after topsoil strip 103958
top of toe 98000
5958

wid
5000
1441
6441

$\frac{6483}{13500} \times 3000$

Fill to structures; non-selected exc. material (E 613)

150.00			Backfilling above toe of retaining wall to stripped level.
6.44			
5.96	5757.36		

Retaining Wall

				Explanatory Notes
				5. FILLED AREA (CONT)
			Ddt	
			Fill to structures; non-selected exc. material a.b.d. (E613)	Adjustment on last item for either end of the filling to the wall which has a sloping bank and is therefore not filled
2/½/	5.96			
	16.00			
	6.44	614.11		

N.B. in actual fact this is very slightly over size but the calculation to compute the actual size would be so complicated that it would never be done in practice.

Retaining Wall

Explanatory Notes

5. FILLED AREA (CONT)

Level	Av. g.l after topsoil strip	
E1	1/103700	103700
E2	2/103700	207400
E3	2/103700	207400
E4	2/103700	207400
E5	2/103700	207400
E6	2/103800	207600
E7	1/103200	103200
F1	2/105200	210400
F2	4/105500	422000
F3	4/105600	422400
F4	4/105800	423200
F5	4/106000	424000
F6	4/105600	422400
F7	2/105200	210400
G1	2/107000	214000
G2	4/107200	428800
G3	4/107600	430400
G4	4/107700	430800
G5	4/107600	430400
G6	4/107500	430000
G7	2/107100	214200
H1	2/108800	217600
H2	4/109000	436000
H3	4/109300	437200
H4	4/109600	438400
H5	4/109700	438800
H6	4/109600	438400
H7	2/109200	218400
I1	1/111000	111000
I2	2/111000	222000
I3	2/111200	222400
I4	2/111300	222600
I5	2/111500	223000
I6	2/111500	223000
I7	1/111400	111400
	96)	10317700
		107476
Less topsoil		100
		107376

Explanatory Notes

5. FILLED AREA (CONT)

Timesing	Dimension	Squaring	Description	Explanatory Notes
			Total quant. of exc. mat. left for fill.	This is the calculation for the total quantity of excavated material still left for filling after backfilling to the structure
			36071·50	
			5757·56	
			Less 614·11 5143·45	
			30928·05 m^3	
			Gen. fill	
			Dep.	
			Fin. lev. 111000	
			topsoil 150	
			110850	
			Lev. of extg grnd after topsoil strip 107376	
			3474	
			Len	
			Len. at lower lev. =	
			$99000 - \left(\frac{111500-103700}{111500-98000} \times 3000\right)$	
			= 97267	
			Av. len. 97267	
			99000	
			2) 196267	
			98134	
	98·13		Gen. fill; imported mat. as spec. Clause 9.1	The precise nature of the material would be given in the Specification. The item description should therefore refer to the clause or give the relevant details.
	100·00		(E635)	
	3·47			
		34051·11		

				Explanatory notes
			5. FILLED AREA (CONT)	
			Dedt Gen fill; imported mat. Spec. Clause 9.1 abd. (E635)	Adjustment on previous item for the banks which are not filled.
2/1/3/1/2/	16.00 3.47 98.13	1816.06		For filling with the excavated material still surplus which reduces the amount of import. (see next item)
	30928.05 1.00 1.00	30928.05		
		32744.11		
			Add Gen. fill; non-selected exc. mat. other than topsoil or rock (E633)	This is the total quantity of excavated material left for general filling.
	30928.05 1.00 1.00			There are no filling or preparation items to be measured as all the filled surfaces are to receive topsoil (M22 and 23)
		30928.05		

Retaining Wall

Timesing	Dimensions	Squaring	Description	Explanatory Notes
			6. TOPSOIL AND LANDSCAPING	
			0.15) 3000 m^3 = 20000 m^2	All the excavated topsoil will be re-used because the total quantity required exceeds the total available. The total volume of excavated topsoil is converted to m^2 of 150mm thick spread material and this is then deducted from the total amount required to give the volume of imported material.
			Fill to 150mm depth; Exc. topsoil (E 641)	
	20000.00 1.00	20000.00		
			Fill to 150mm depth; imp. topsoil (E 642)	
	100.00 150.00	15000.00	(cut area)	
	99.00 150.00	14850.00	(fill area)	
		29850.00		
			Ddt Ditto (E 642)	The deduction for the sloping areas is because the CESMM requires that filling to a constant thickness be measured according to its angle of inclination to the horizontal
2/½/	100.00 16.00	1600.00	(for sloping banks)	
2/½/	99.00 16.00	1584.00	(for sloping banks)	
	20000.00 1.00	20000.00	(for exc. mat)	
		23184.00		

Retaining Wall

Explanatory Notes

6. TOPSOIL AND LANDSCAPING (CONT)

Times	Dimensions	Quantity	Description	Explanatory Notes
			cut side base length: $\sqrt{3.60^2 + 16.00^2}$ = 16.40	The angle of inclination is most easily calculated either by the use of logarithms or by scaling as shown.
			fill side base length: $\sqrt{3.47^2 + 16.00^2}$ = 16.37	
			Add	
2/½/	101.27 16.40	1660.83	Fill; 150mm dep; imp. topsoil; 10-45° to the horiz. (E 642.1)	
2/½/	100.28 16.40	1644.59		
		3305.42	& Seeding with low maintenance grass seed as Spec. Clause 2.13; over 10° to horiz. (E830)	The sloping areas can be treated as triangles for the purposes of measurement. The area is slightly undermeasured, but is considered sufficiently accurate for this purpose
	100.00 150.00	15000.00	Seeding as Spec. Clause 2.12; n.e 10° to the horiz. (E 830.1)	The precise method, types, and laying rates would be contained in the specification to which reference should be made; item coverage should define exactly what treatments are to be included in the rates
	99.00 150.00	14850.00		
		29850.00		

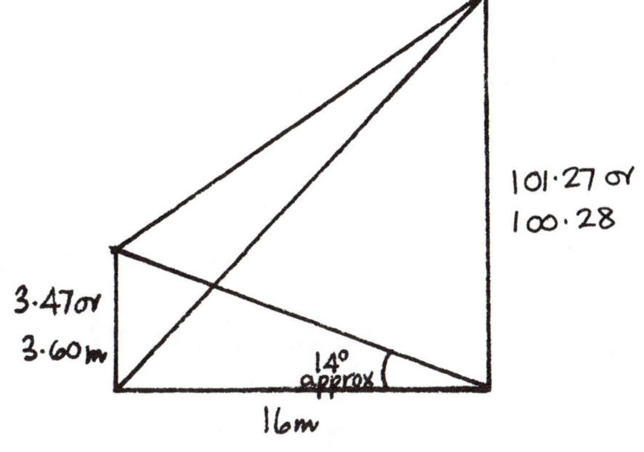

Retaining Wall

Explanatory Notes

6. TOPSOIL AND LANDSCAPING (CONT)

Timesing	Dimensions	Squaring	Description	Explanatory Notes
			Ddt Seeding as Spec. Clause 2.12; n.e. 10° to the horiz. a.b.d. (E 830.1)	
2/½/	100.00 16.00	1600.00	(for sloping banks measured elsewhere)	
2/½/	99.00 16.00	1584.00		
		3184.00		
	3305	3305	Plant shrubs; Acer griseum 45-60 cms high (to sloping banks 1/m²) (E 850)	Stating that trees are bare root is not a CESMM requirement but is an additional description under 5.10.
	24	24	Plant trees; Acer campestre; heavy standard; bare root (E 860)	The specification will normally detail the planting, treatment and techniques required, and it may be necessary to refer to these in the item description; item coverage rules should be extended or included to state exactly what is to be allowed for in the rates.
	48	48	Plant trees; Acer platanoides; feathered; bare root (E860.1)	

Chapter 6
ROAD

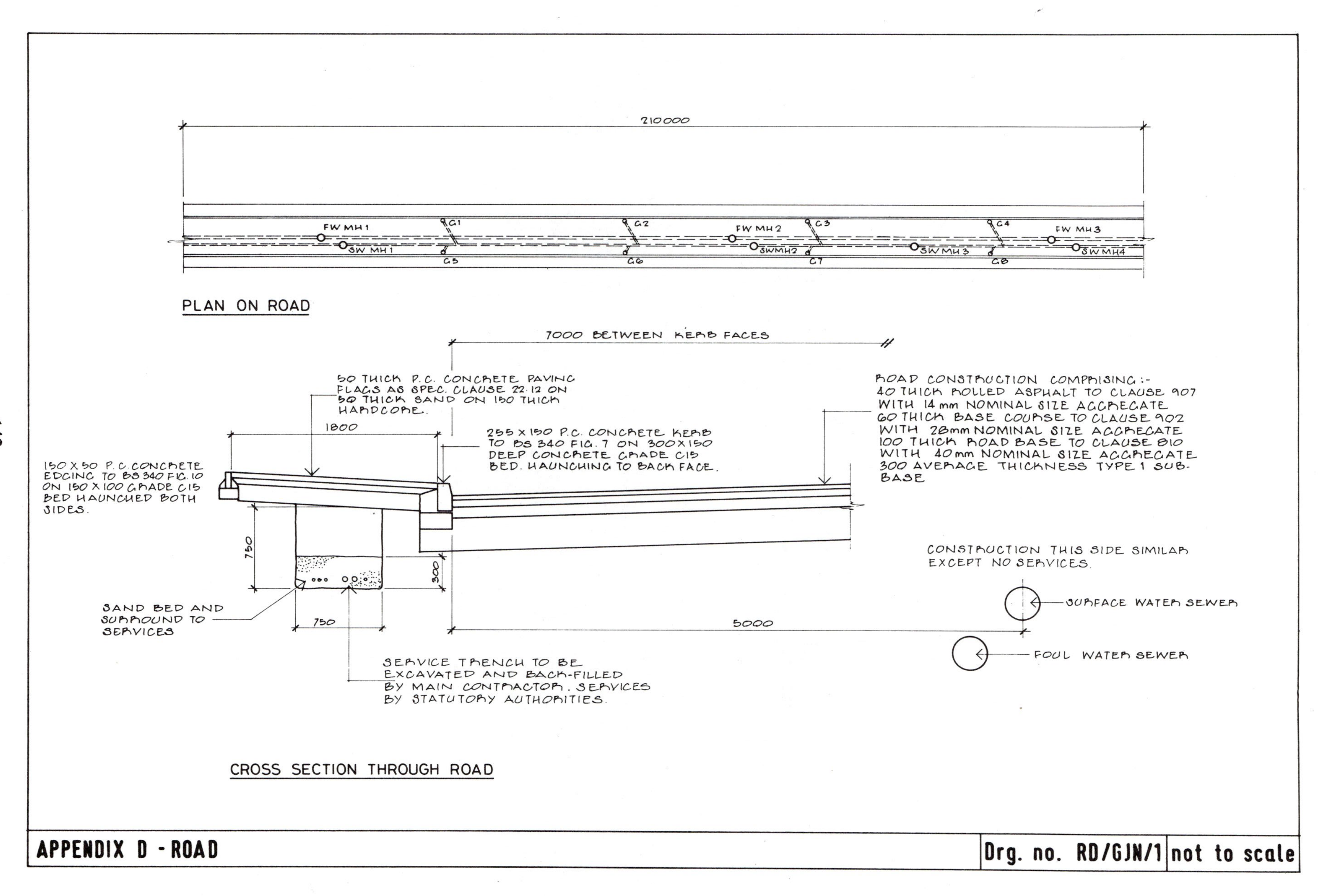

210000
FW MH1
SW MH1
C1
C2
C3
C4
C5
C6
C7
C8
FW MH2
SWMH2
SW MH3
FW MH3
SW MH4
PLAN ON ROAD
7000 BETWEEN KERB FACES
50 THICK P.C. CONCRETE PAVING FLAGS AS SPEC. CLAUSE 22.12 ON 50 THICK SAND ON 150 THICK HARDCORE.
1800
255 X 150 P.C. CONCRETE KERB TO BS 340 FIG. 7 ON 300 X 150 DEEP CONCRETE GRADE C15 BED. HAUNCHING TO BACK FACE.
ROAD CONSTRUCTION COMPRISING:- 40 THICK ROLLED ASPHALT TO CLAUSE 907 WITH 14 mm NOMINAL SIZE AGGREGATE 60 THICK BASE COURSE TO CLAUSE 902 WITH 28mm NOMINAL SIZE AGGREGATE 100 THICK ROAD BASE TO CLAUSE 810 WITH 40mm NOMINAL SIZE AGGREGATE 300 AVERAGE THICKNESS TYPE 1 SUB-BASE
150 X 50 P.C. CONCRETE EDGING TO BS 340 FIG. 10 ON 150 X 100 GRADE C15 BED HAUNCHED BOTH SIDES.
750
300
SAND BED AND SURROUND TO SERVICES
750
5000
CONSTRUCTION THIS SIDE SIMILAR EXCEPT NO SERVICES.
SURFACE WATER SEWER
FOUL WATER SEWER
SERVICE TRENCH TO BE EXCAVATED AND BACK-FILLED BY MAIN CONTRACTOR. SERVICES BY STATUTORY AUTHORITIES.
CROSS SECTION THROUGH ROAD
APPENDIX D - ROAD
Drg. no. RD/GJN/1
not to scale

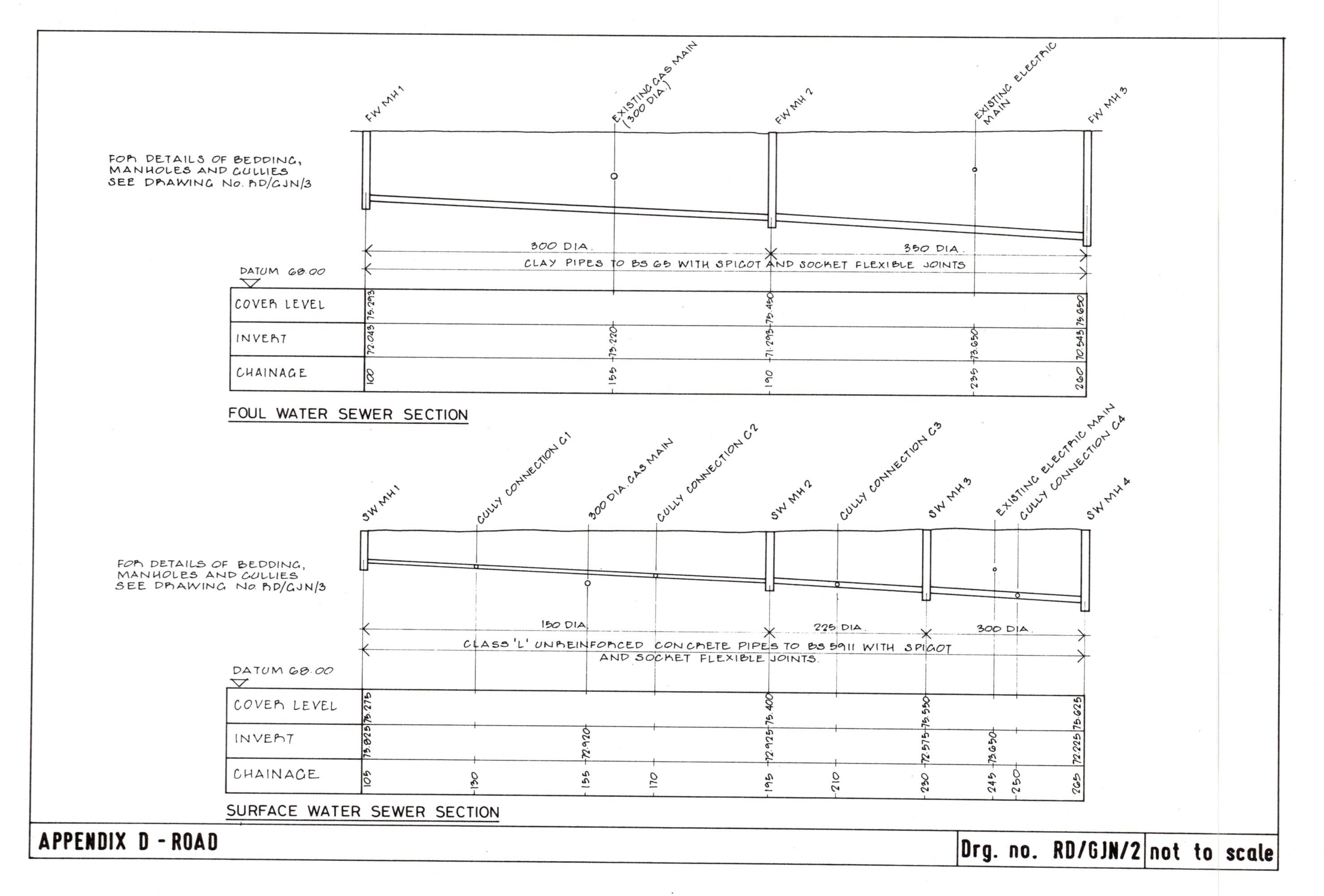

FW MH1
EXISTING GAS MAIN (300 DIA.)
FW MH 2
EXISTING ELECTRIC MAIN
FW MH 3
FOR DETAILS OF BEDDING, MANHOLES AND GULLIES SEE DRAWING No. RD/GJN/3
300 DIA.
350 DIA.
CLAY PIPES TO BS 65 WITH SPIGOT AND SOCKET FLEXIBLE JOINTS
DATUM 68.00
COVER LEVEL
INVERT
CHAINAGE
75.293
72.043
100
73.220
155
75.450
71.293
190
73.650
235
73.650
70.543
260
FOUL WATER SEWER SECTION
SW MH1
GULLY CONNECTION C1
300 DIA. GAS MAIN
GULLY CONNECTION C2
SW MH 2
GULLY CONNECTION C3
SW MH3
EXISTING ELECTRIC MAIN
GULLY CONNECTION C4
SW MH 4
FOR DETAILS OF BEDDING, MANHOLES AND GULLIES SEE DRAWING No. RD/GJN/3
150 DIA.
225 DIA.
300 DIA.
CLASS 'L' UNREINFORCED CONCRETE PIPES TO BS 5911 WITH SPIGOT AND SOCKET FLEXIBLE JOINTS.
DATUM 68.00
COVER LEVEL
INVERT
CHAINAGE
75.275
73.825
105
130
72.920
155
170
75.400
72.925
195
210
75.550
72.575
230
73.650
245
250
73.625
72.225
265
SURFACE WATER SEWER SECTION
APPENDIX D - ROAD
Drg. no. RD/GJN/2
not to scale

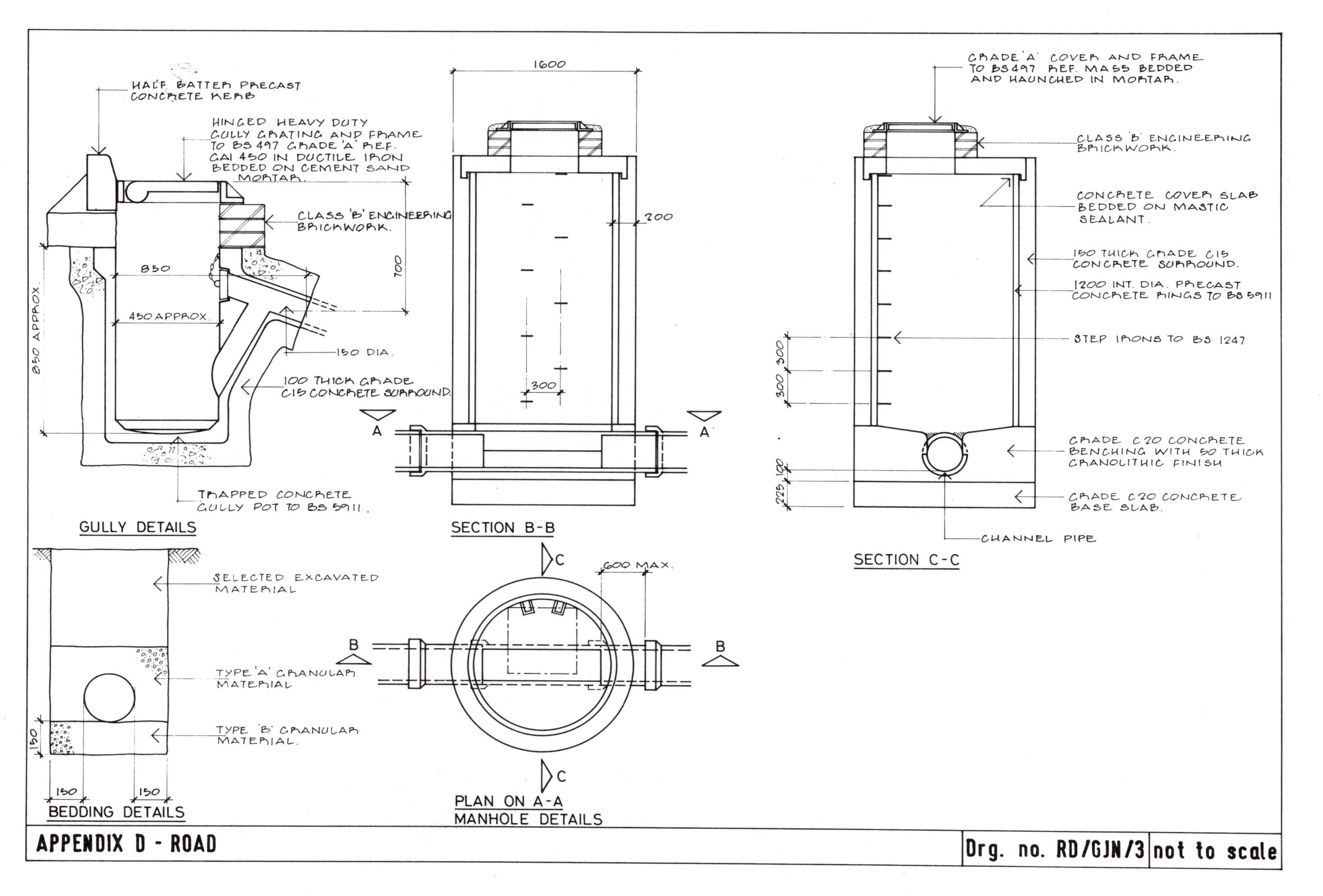
HALF BATTER PRECAST CONCRETE KERB
HINGED HEAVY DUTY GULLY GRATING AND FRAME TO BS 497 GRADE 'A' REF. GAI 450 IN DUCTILE IRON BEDDED ON CEMENT SAND MORTAR.
CLASS 'B' ENGINEERING BRICKWORK.
850
450 APPROX.
700
850 APPROX.
150 DIA.
100 THICK GRADE C15 CONCRETE SURROUND.
TRAPPED CONCRETE GULLY POT TO BS 5911.
GULLY DETAILS
1600
200
300
A
A
SECTION B-B
GRADE 'A' COVER AND FRAME TO BS497 REF. MA55 BEDDED AND HAUNCHED IN MORTAR.
CLASS 'B' ENGINEERING BRICKWORK.
CONCRETE COVER SLAB BEDDED ON MASTIC SEALANT.
150 THICK GRADE C15 CONCRETE SURROUND.
1200 INT. DIA. PRECAST CONCRETE RINGS TO BS 5911
STEP IRONS TO BS 1247
300
300
GRADE C20 CONCRETE BENCHING WITH 50 THICK GRANOLITHIC FINISH
100
225
GRADE C20 CONCRETE BASE SLAB.
CHANNEL PIPE
SECTION C-C
SELECTED EXCAVATED MATERIAL
TYPE 'A' GRANULAR MATERIAL
TYPE 'B' GRANULAR MATERIAL.
150
150
150
BEDDING DETAILS
C
600 MAX.
B
B
C
PLAN ON A-A
MANHOLE DETAILS
APPENDIX D - ROAD
Drg. no. RD/GJN/3
not to scale

Explanatory Notes

Drawing Numbers RD/GJN/1

RD/GJN/2

RD/GJN/3

The following example show a length of road and footpath with associated drainage and services.

The example demonstrates measurement in accordance with Classes I, J, K, L and R only. Rock head is at level 71750.
The foulwater pipework is measured between manholes FWMH1 and FWMH3 and the surface water pipework between SWMH1 and SHMH4.
In practice the pipework would be scheduled in order to simplify the taking off process and this is shown in the example, demonstrating how each length of pipework is calculated within each of the depth classifications.

FOUL WATER PIPEWORK				DEPTH CLASSIFICATION			
RUN	TYPE	DIA	2·5-3	3-3·5	3·5-4	4-4·5	4·5-5
FWMH1-MH2	CLAY	300	24·48	48·95	15·37		
CALCS							
1ST dep.	75293 72043 3250		Length:	Length:	Length:		
Less road	500 2750		$\frac{3-2.75}{0.907} \times 88.80$	$\frac{3.5-3}{0.907} \times 88.80$	$\frac{3.657-3.5}{0.907} \times 88.80$		
			= 24·48	= 48·95	= 15·37		
2ND dep.	75450 71293 4157						
Less road	500 3657						
total dep. of fall	3657 2750 907						
Total pipe length	190000 100000 90000			The calculation of the pipe length within each depth classification is done by interpolation. The total pipe length is measured from the inside faces of manholes (M5)			
MH 2/½/1600	1600 88400						
Walls 2/200	400 88800						

FOUL WATER PIPEWORK				DEPTH CLASSIFICATION			
RUN	TYPE	DIA	2·5-3	3-3·5	3·5-4	4-4·5	4·5-5
FWMH 2-MH3	CLAY	350			24·84	36·21	7·75
CALCS							
1ST dep.	75450				Length:	Length:	Length:
	71 293				$\frac{4-3{\cdot}657}{0{\cdot}95} \times 68{\cdot}80$	$\frac{4{\cdot}5-4}{0{\cdot}95} \times 68{\cdot}80$	$\frac{4{\cdot}607-4{\cdot}5}{0{\cdot}95} \times 68{\cdot}80$
	4157						
Less road	500				= 24·84	= 36·21	= 7·75
	3657						
2ND dep.	75650						
	70543						
	5107						
Less road	500						
	4607						
total dep. of fall	4607						
	3657						
	950						
Total pipe length	260 000						
	190 000						
	70 000						
MH 2/½/1600	1 600						
	68 400						
Walls 2/200	400						
	68 800						

SURFACE WATER PIPEWORK				DEPTH CLASSIFICATION			
RUN	TYPE	DIA	n.e. 1·5	1·5-2	2-2·5	2·5-3	3-3·5
SWMH 1-MH2	CONC	150	47·65	41·15			
CALCS							
1ST dep.	75275		Length:	Length:			
	73825		$\frac{1.5-0.95}{1.025} \times 88.80$	$\frac{1.975-1.5}{1.025} \times 88.80$			
	1450		= 47·65	= 41·15			
Less road	500						
	950						
2ND dep.	75400						
	72925						
	2475						
Less road	500						
	1975						
total dep. of fall	1975						
	950						
	1025						
Total pipe length	195000						
	105000						
	90 000						
MH 2/½/1600	1 600						
	88 400						
Walls 2/200	400						
	88 800						

SURFACE WATER PIPEWORK / DEPTH CLASSIFICATION

RUN	TYPE	DIA	n.e. 1·5	1·5 - 2	2 - 2·5	2·5 - 3	3 - 3·5
SWMH 2-MH3	CONC	225		1·69	32·11		
CALCS				Length:	Length:		
1st dep.	75400 72925 2475			$\frac{2-1\cdot975}{0\cdot50} \times 33\cdot80$	$\frac{2\cdot475-2}{0\cdot50} \times 33\cdot80$		
Less road	500 1975			= 1·69	= 32·11		
2nd dep.	75550 72575 2975						
Less road	500 2475						
total dep. of fall	2475 1975 500						
Total pipe length	230000 195000 35000						
MH 2/½/1600	1600 33400						
Walls 2/200	400 33800						

SURFACE WATER		PIPEWORK		DEPTH CLASSIFICATION			
RUN	TYPE	DIA	n.e. 1·5	1·5 – 2	2 – 2·5	2·5 – 3	3 – 3·5
SWMH 3 – MH4	CONC	300			1·99	31·81	
CALCS							
1st dep.	75550 72575 2975				Length:	Length:	
Less road	500 2475				$\frac{2\cdot5-2\cdot475}{0\cdot425} \times 33\cdot80$ = 1·99	$\frac{2\cdot9-2\cdot5}{0\cdot425} \times 33\cdot80$ = 31·81	
2nd dep.	75625 72225 3400						
Less road	500 2900						
total dep. of fall	2900 2475 425						
Total pipe length	265000 230000 35000						
MH 2/½/1600	1600 33400						
Walls 2/200	400 33800						

SURFACE WATER DRAINAGE - GULLY CONNECTIONS

- N.B. it is assumed that the inverts of adjacent gully connections are the same because there will be very little difference between them as they are so close together.

depth at gully

	invert from wearing course		700
Less	road construction	40	
		60	
		100	
		300	500
			200

commencing surface levels at sewer

G1/G5 75400 − 75275 × $\frac{24.4}{88.8}$ + 75275 = 75309
Less road 500
74809

G2/G6 75400 − 75275 × $\frac{64.4}{88.8}$ + 75275 = 75366
Less road 500
74866

G3/G7 75550 − 75400 × $\frac{14.4}{33.8}$ + 75400 = 75464
Less road 500
74964

G4/G8 75625 − 75550 × $\frac{19.4}{33.8}$ + 75550 = 75593
Less road 500
75093

Inverts at sewer

G1/G5 73825 − ($\frac{24.4}{88.8}$ × 1025) = 73543

G2/G6 73825 − ($\frac{64.4}{88.8}$ × 1025) = 73082

G3/G7 72925 − ($\frac{14.4}{33.8}$ × 500) = 72712

G4/G8 72575 − ($\frac{19.4}{33.8}$ × 425) = 72331

Depths at sewer

G1/G5 74809 − 73543 = 1266

G2/G6 74866 − 73082 = 1784

G3/G7 74964 − 72712 = 2252

G4/G8 75093 − 72331 = 2762

Pipe lengths

G1, G2, G3, G4. 5000
gully 850
4150

G5, G6, G7, G8. 7000
5000
2000
gully 850
1150

SURFACE WATER DRAINAGE – GULLY CONNECTIONS

GULLY RUN	TYPE	DIA	DEPTH CLASSIFICATION n.e. 1·5	1·5 – 2	2 – 2·5	2·5 – 3
G1/G5	CONC	150	1·15 4·15			
G2/G6	CONC	150	$\frac{1·5-0·2}{1·784-0·2} \times 4·15 \times 1·15$ = 3·41 = 0·94	$\frac{1·784-1·5}{1·784-0·2} \times 4·15 \times 1·15$ = 0·74 = 0·21		
G3/G7	CONC	150	$\frac{1·5-0·2}{2·252-0·2} \times 4·15 \times 1·15$ = 2·63 = 0·73	$\frac{2·0-1·5}{2·252-0·2} \times 4·15 \times 1·15$ = 1·01 = 0·28	$\frac{2·252-2·0}{2·252-0·2} \times 4·15 \times 1·15$ = 0·51 = 0·14	
G4/G8	CONC	150	$\frac{1·5-0·2}{2·762-0·2} \times 4·15 \times 1·15$ = 2·11 = 0·58	$\frac{2·0-1·5}{2·762-0·2} \times 4·15 \times 1·15$ = 0·81 = 0·22	$\frac{2·5-2·0}{2·762-0·2} \times 4·15 \times 1·15$ = 0·81 = 0·22	$\frac{2·762-2·5}{2·762-0·2} \times 4·15 \times 1·16$ = 0·42 = 0·12

			FOUL WATER DRAINAGE
			PIPES; FWMH 1-FWMH3
			Vitrified clay pipes; B.S. 65 normal quality; spigot & socket flexible joints; Comm. Surf underside of road construction
			Nom. bore 300mm; in tr. dep. 2·5-3m (I 125)
	24·48		(FWMH 1-MH2)
		24·48	
			Nom. bore 300mm; in tr. dep 3-3·5m (I 126)
	48·95		(FWMH 1-MH2)
		48·95	
			Nom. bore 300mm; in tr. dep 3·5-4m (I 127)
	15·37		(FWMH 1-MH2)
		15·37	

Explanatory Notes

It is normal to measure and bill pipework separately according to its function. In this case this means measuring the foul and surface water separately

A1 requires the location of the pipework to be stated and in this case it forms part of the general heading. The Commencing Surface is stated as it is not also the Original Surface (A4)

				Explanatory Notes
			FOUL WATER (CONT)	
			PIPES; FWMH1-MH3 (CONT)	
			Vitrified clay pipes (cont)	
			Nom. bore 350mm; in tr. dep 3·5-4m (I 137)	
	24·84	24·84	(FWMH 2-MH3)	
			Nom. bore 350mm; in tr. dep 4-4·5m (I 138)	Trenches over 4m deep must be stated in increments of 0·5m (A6)
	36·21	36·21	(FWMH 2-MH3)	
			Nom. bore 350mm; in tr. dep. 4·5-5m (I 138·1)	
	7·75	7·75	(FWMH2-MH3)	

Explanatory Notes

		Description	Explanatory Notes
		FOUL WATER (CONT)	
		MANHOLES AND PIPEWORK ANCILLARIES	
		Precast conc. manholes with in-situ surr.; as drwg no. RD/GJN/3; cover & fr. to B.S. 497 grade A ref. MA 55	A1 requires that the type or mark numbers be given in the item descriptions. This is done by giving their actual reference. The type and loading duty must be stated (A2) and this is given in the general heading. As an alternative to enumerating the manholes they could have been measured in detail in accordance with other classes of CESMM (see note at bottom of page 55)
		dep 75293 72043 3250 100 3350	
1	1	dep. 3-3.5m; FWMH1 (K155)	
		dep 75450 71293 4157 100 4257	
1	1	dep. 4.26m; FWMH2 (K157)	The actual depth of the manhole must be stated where this exceeds 4m. It is recommended that this be kept to 2 places of decimals. The depth of the manholes is measured to the top of the base slab, which in this case is 100mm below the invert (D2)
		dep 75650 70543 5107 100 5207	
1	1	dep. 5.21m; FWMH3 (K157.1)	

				Explanatory Notes
			FOUL WATER (CONT)	
			MHS. AND PIPEWK. ANCILL. (CONT)	
			Crossings	
	1	1	300 dia. gas main, pipe bore n.e. 300mm (K681) (FWMH1-MH2)	Although not a specific requirement, the diameter of the gas main is stated in order to assist the estimator.
	1		Electric main; pipe bore 300-900mm (K682) (FWMH 2-MH3)	
			PIPEWORK-SUPPORTS	
			FWMH1 72043 FWMH2 71293 750 Rockhead 71750 71293 457 ∴ rock meets invert at chainage $190 - \left(90 \times \frac{457}{750}\right)$ = 135.16	In order to calculate the quantity of rock excavation, the point at which the rock head crosses the invert of the pipe must be determined. This will give the total length of trench in rock. It must be borne in mind that the inverts on the section are pipe inverts, and that there is also a 150mm bed underneath the pipe which must be accounted for.

55.63

				Explanatory Notes
			FOUL WATER (CONT)	
			PIPEWORK-SUPPORTS (CONT)	
			Total pipe length FWMH1 -MH2 in rock 190000 135160 54840 Less ½ × FWMH2 800 54040	Although the pipes are measured to the inside face of the manhole walls, the volumes of extras to excavation are based on the trench length which is shorter by the width of the manhole walls. Extras to excavation must distinguish between those in trenches and those in manholes, and the length of the manhole is therefore deducted from the pipe length.
			Av. dep. FWMH1-MH2 in rock 1st. 0000 2nd. 71750 71293 457 2) 457 229	
			Nom. tr. wid. pipe 300 500 800	The nominal trench width is 500mm greater than the nominal bores of the pipes (D1).
			Total pipe length FWMH 2-MH3 in rock 260000 190000 70000 Less ½ × FWMH2 800 ½ × FWMH3 800 1600 68400	
			Av. dep. FWMH 2-MH3 in rock 1st. 71750 71293 457 2nd 71750 70543 1207 2) 1664 832	
			Nom. tr. wid. pipe 350 500 850	

Explanatory Notes

FOUL WATER (CONT)

PIPEWORK SUPPORTS (CONT)

		Extras to exc. in pipe tr.; exc. of rock (L111)	All the foregoing calculations are based on the invert of the pipe. The volume occupied by the bed underneath the pipe is also in rock and is included in the dimensions separately.
54.04 0.80 0.23	9.94	(FWMH1-MH2)	
54.04 0.80 0.15	6.48	(ditto for bed)	
68.40 0.85 0.83	48.26	(FWMH2-MH3)	
68.40 0.85 0.15	8.72	(ditto for bed)	
	73.40		

		dep
rock level		71750
invert		71293
		457
base	100 200	300
		757
rock level		71750
invert		70543
		1207
base	100 200	300
		1507

The depth of rock excavation in manholes must take into account the depth of the base below the invert.

				Explanatory Notes
			FOUL WATER (CONT)	
			PIPEWORK SUPPORTS (CONT)	
			Extras to exc. in mhs; exc. of rock (L121)	
π/	0·80 0·80 0·76	1·53	(FWMH2)	
π/	0·80 0·80 1·51	3·04	(FWMH3)	
			Beds 150 dp in type B gran. mat.; with surr. in type A gran. mat.	Because the surround is in a different material to the bed, the item description must so state. The actual nominal bore of the pipe is given because this is considered to be more helpful to the estimator than the ranges given in the Third Division. Note that the total length of the bed is slightly shorter than the pipe length because of the manhole walls.
			Len 190000 100000 90000 2/½/manhole 1600 88400	
			pipe nom. bore 300 mm (L332)	
	88·40		(FWMH1-MH2)	
		88·40	Len 260000 190000 70000 2/½/manhole 1600 68400	
			pipe nom. bore 350 mm (L333)	
	68·40		(FWMH2-MH3)	
		68·40		

			Explanatory Notes
		SURFACE WATER DRAINAGE	
		PIPES; SWMH 1 - SWMH4	
		Precast conc. pipes and fittings; B.S. 5911 class L unreinforced; spigot & socket flexible joints; Comm. surf. underside of road construction	In order to prevent undue repetition, the fittings to the surface water pipework are included with the general measurements of the pipes rather than being classified under a seperate heading.
		Nom. bore 150mm; in. tr. dep. n.e. 1.5m (I 212)	
47.65		(SWMH 1-MH2)	
	47.65		
		Nom. bore 150mm; in. tr. dep. 1.5-2m (I 213)	
41.15		(SWMH 1- MH2)	
	41.15		
		Branches; 150×150 ×150mm (J 221)	
2/ 1	2	(gully connections G.1 G2	
2/ 1	2	G.7 G.8)	
	4		
		Nom. bore 225mm; in tr. dep 1.5-2m (I 223)	
1.69		SWMH 2-MH 3	
	1.69		

Road

				Explanatory Notes
			SURFACE WATER (CONT)	
			PIPES; SWMH1 - MH4 (CONT)	
			Precast conc. pipes & fittings (cont)	
			Nom. bore 225mm; in tr. dep. 2-2·5m (I224)	
	32·11	32·11	(SWMH 2-MH3)	
			Branches; 225 x 225 x 150mm (J222)	
2/	1	2	(gully connections G.3. G7)	
		2		
			Nom. bore 300mm; in tr. dep. 2-2·5m (I224.1)	
	1·99	1·99	(SWMH 3 - MH4)	
			Nom. bore 300mm; in tr. dep. 2·5-3m (I225)	
	31·81	31·81	(SWMH 3-MH4)	
			Branches; 300 x 300 x 150 mm (J222.1)	
2/	1	2	(gully connection G4. G8)	
		2		

			Explanatory Notes
		SURFACE WATER (CONT)	
		PIPES; SWMH 1-MH4 (CONT)	
		Precast conc. pipes & fittings (cont)	
		Nom. bore 150mm; in tr. dep. n.e 1·5m (I 212)	The length of the gully connections is measured from the centreline of the sewer (M3). Although not specifically stated the pipe is not measured through the gully itself as this is not a fitting (M3)
1·15		G5 (gully connections)	
4·15		G1	
3·41		G6	
0·94		G2	
2·63		G7	
0·73		G3	
2·11		G8	
0·58		G4	
	15·70		
		Nom. bore 150mm; in tr. dep. 1·5-2m (I 213)	
0·74		G6 (gully connections)	
0·21		G2	
1·01		G7	
0·28		G3	
0·81		G4	
0·22		G8	
	3·27		

Road

				Explanatory Notes
			SURFACE WATER (CONT)	
			PIPES; SWMH1-MH4 (CONT)	
			Precast conc. pipes & fittings (cont)	
			Nom. bore 150mm; in tr. dep. 2-2.5m (I 2.14)	
	0.51		G7 (gully connections)	
	0.14		G3	
	0.81		G8	
	0.22		G4	
		1.68		
			Nom. bore 150mm; in tr. dep. 2.5-3m (I 2.15)	
	0.42		G8 (gully connections)	
	0.12		G4	
		0.54		

Road

				Explanatory Notes
			SURFACE WATER (CONT)	
			MANHOLES AND PIPEWORK ANCILLARIES	
			Precast conc. manholes with in-situ surr; as drwg no. RD/GJN/3; cover & fr. to BS 497 grade A ref MA55	
			dep 75275 73825 1450 100 1550	
1			dep. 1·5–2m; SWMH1 (K152)	
	1			
			dep 75400 72925 2475 100 2575	
1			dep. 2·5–3m; SWMH2 (K154)	
	1			
			dep 75550 72575 2975 100 3075	
1			dep. 3–3·5m; SWMH3 (K155)	
	1			

Road

				Explanatory Notes
			SURFACE WATER (CONT)	
			MHs. AND. PIPEWORK ANCILL (CONT)	
			Precast conc. mhs (conts)	
			dep 75625 72225 3400 100 3500	
			dep. 3-3·5m; SWMH4	
			(K155)	
	1			
		1		
			Gullies	
			Precast conc. trapped; as drwg. no. RD/GJN/3; cover & fr. to B.S. 497 grade A ref. GA1450	The type and loading duty of gully covers must be stated in item descriptions
			(K360)	
	8/ 1		(all gullies)	
		8		

Road

				Explanatory Notes
			SURFACE WATER (CONT)	
			MHS AND PIPEWK ANCILL (CONT)	
			Crossings	Note that the surface water sewer passes above the gas main and does not have to be measured as a crossing.
			Electric main; pipe bore n.e. 300mm (K681.1)	
	1	1		Also note that the lowest invert of the pipe including its bed (72075) or manhole to the underside of its base (SWMH4 at 71925) are both higher than the level of rock and hence there will not be any rock excavation for surface water drainage.
			PIPEWORK SUPPORTS	
			Beds 150dp in type B gran. mat; with surr. in type A gran. mat.	Note that the length of the bed and surround is measured along the pipe centre line (M11). Although not specifically stated, the length occupied by the gully is deducted as this is not classified as a fitting or valve.
			pipe nom. bore 150mm (L331)	
	88.40	88.40	(SWMH 1 - MH2)	
4/	4.15	16.60	(gully runs)	
4/	1.15	4.60		
		109.60		

<u>Road</u>

				Explanatory Notes
			SURFACE WATER (CONT)	
			PIPEWORK SUPPORTS (CONT)	
			Beds 150dp in type B gran mat (cont)	
			pipe nom. bore 225mm (L332.1)	
	33.40		(SWMH 2-MH3)	
		33.40		
			pipe nom. bore 300mm (L332.2)	
	33.40		(SWMH 3-MH4)	
		33.40		

			Explanatory Notes
		SERVICE TRENCH FOR STATUTORY AUTHORITIES	
		PIPEWORK - MANHOLES	
		Tr. for pipes or cables not laid by contractor; backfilling above bed & surr. with selected Exc. Mat. after install. of services by others	Although not specifically required, details of backfilling requirements (if any) should always be given in item descriptions for service trenches
		Cross-sect. area 0.5-0.75m² (K483)	
210.00	210.00		
		PIPEWORK SUPPORTS	This is a rogue item as it is not classified strictly in accordance with the Third Division, and is therefore coded 9. Also, the size of the bed and surround has been stated because it is not particularly related to the bore of the services or pipes.
		Sand bed & surr. 750 x 300mm around services (L3&419)	
210.00	210.00		

Road

				Explanatory Notes
			ROADS AND PAVINGS	
			Sub-bases, flexible road bases and surfacing.	
	210.00 7.60	1596.00	Gran. mat. D.Tp. specified type 1; dep. 300mm (R117) Wid 7000 Kerb fndn 2/300 600 7600 (sub base)	M1 states the width of each course to be measured at the top surface. In the case of the sub base this is taken as extending under the kerb foundation. It is not necessary to state the depth in bands as the actual depth is given (Paragraph 3.10)
	210.00 7.00	1470.00	Dense tarmacadam; D.Tp. clause 810; dep. 100mm (R253) (road base) &	It is not necessary to state the size of aggregate as there there is only one size in the Specification. No deductions are made for the areas of manholes and gullies as there plan size does not exceed $1m^2$ (M1).
			Rolled asphalt; D.Tp. clause 902; 28mm nom. size agg.; dep. 60mm (R322) (base course) &	It is necessary to state the aggregate size for both the base course and wearing course as there are several sizes in the Specification
			Rolled asphalt; D.Tp. clause 907; 14mm nom. size agg.; dep. 40mm (R322.1) (wearing course)	

				Explanatory Notes
			ROADS AND PAVINGS (CONT)	
			Kerbs, channels and edgings	
			Precast. conc. Kerb to BS.7263: Part 1, Fig 1(F); straight; incl. fndn & haunch as drwg no. RD/GJN/1 (R631)	It is not necessary to state the dimensions in the item description in accordance with A7 as all the relevant information is contained on the drawing.
2/	210.00	420.00	&	
			Precast conc. Edging to BS.7263 Part 1, Fig (lm); straight; incl. fndn & haunch. (R661)	The length of the edging is exactly the same as the kerb, hence the use of the ampersand.

Road

				Explanatory Notes
			ROADS AND PAVINGS (CONT)	
			Light duty pavements	
			Hardcore base; dep. 150mm (R724) wid 1800 Kerb haunch 300-150 = 150 Edging haunch $\frac{150-50}{2}$ = 50 200 1600	As the width is defined as the width along the top surface. the haunch to both the kerb and edging must be deducted.
2/	210.00 1.60		(footpath)	
		672.00		
			Sand base (R712)	Although the bottom of the sand layer is the same as the hardcore width the top surface is wider and it is this width upon which the measurement is based.
2/	210.00 1.80		(blinding)	
		756.00	&	
			Precast conc. flags as spec. clause 22.12 (R782)	

Chapter 7
REINFORCEMENT

REV.

BAR SCHED. REF. BBS 21 A

DATE JAN. 1986

MEMBER	BAR MARK	TYPE & SIZE	No. OF MBRS.	No. IN EACH	TOTAL No.	LENGTH OF EACH BAR	SHAPE CODE	A	B	C	D	E/r
COMPONENT 'A'	24	T16	2	24	48	1450	37	1290				
	25	T10	1	86	86	1675	37	1205				
	26	T10	3	108	324	5075	20	STRAIGHT				
	27	T12	1	8	8	3000	38	1370	330			
COMPONENT 'B'	21	T10	2	20	40	5550	37	5385				
	22	T10	1	72	72	5425	37	5260				
	23	T12	6	36	216	2150	38	1000	215			
	24a	T12	1	4	4	2675	37	2515				
	24b	T12	10	4	40	2175	37	2010				
	24c	T10	5	8	40	1925	37	1755				
	24d	T16	1	8	8	1750	37	1580				
	24e	T16	8	8	64	1625	37	1455				
	24f	T10	1	24	24	1450	37	1290				
	25a	T10	10	78	780	1675	37	1205				
	26a	T20	1	100	100	5075	20	STRAIGHT				
	27a	T20	4	4	16	3000	38	1370	330			
	28	T10	4	8	32	2600	20	STRAIGHT				
	29	T10	4	8	32	1225	37	1055				
	30	T12	6	32	192	4350	38	2050	330			
	31a	T16	2	4	8	2075	38	920	330			

ALL BENDING DIMENSIONS ARE IN ACCORDANCE WITH BS 4466

Explanatory Notes

Bar bending schedule BBS-21

Reinforcement quantities may be prepared by one of three methods.

1) By direct taking off from the drawings.

2) From bar bending schedules

3) From average weights of steel per m^3 of concrete, divided into various diameters. This method should only be used for the tender quantities and never for the final account.

The following is an example of the calculation of reinforcement weights from a typical bar bending schedule.
The schedule gives the total number and girth of each bar as calculated in accordance with BS4449.
The measurement involves abstracting the bar lengths according to their diameter and type, multiplying them by the total number of each and applying the conversion factor to arrive at the gross weight.
The schedules are normally prepared by the Engineer.

Reinforcement

Explanatory Notes

Deformed High Yield steel bars to BS. 4449

10mm nominal size (G 523)

Timesing	Dimension	Squaring	Description
			(bar mark No)
86/	1.68	144.48	(25
324/	5.08	1645.92	(26
40/	5.55	222.00	(21
72/	5.43	390.96	(22
40/	1.93	77.20	(24b
24/	1.45	34.80	(24f
780/	1.68	1310.40	(25
32/	2.60	83.20	(28
32/	1.23	39.36	(29
		3948.32	
		x0.616	kg/m
		2432.17	÷ 1000 = 2.43217T

The weights of reinforcement per mm² of cross-sectional area are given in kg/m in the British Standard. The total weight of reinforcement must therefore be divided by 1000 to convert it to tonnes.

12mm nominal size (G524)

Timesing	Dimension	Squaring	Description
			(bar mark No)
8/	3.00	24.00	(27
216/	2.15	464.40	(23
4/	2.68	10.72	(24a
40/	2.18	87.20	(24b
192/	4.35	835.20	(30
		1421.52	
		x0.888	kg/m
		1262.31	÷ 1000 = 1.26231T

Reinforcements

Explanatory Notes

			16mm nominal size (G 525)
			(bar mark No)
48/	1.45	69.60	(24
8/	1.75	14.00	(24d
64/	1.63	104.32	(24e
8/	2.08	16.64	(31a
		204.56	
		x1.579	kg/m
		323.00	÷ 1000 = 0.3230T
			20mm nominal size (G526)
			(bar mark No.)
100/	5.08	508.00	(26a
16/	3.00	48.00	(27a
		556.00	
		x 2.466	kg/m
		1371.10	÷ 1000 = 1.3711 T

Chapter 8
GATEHOUSE

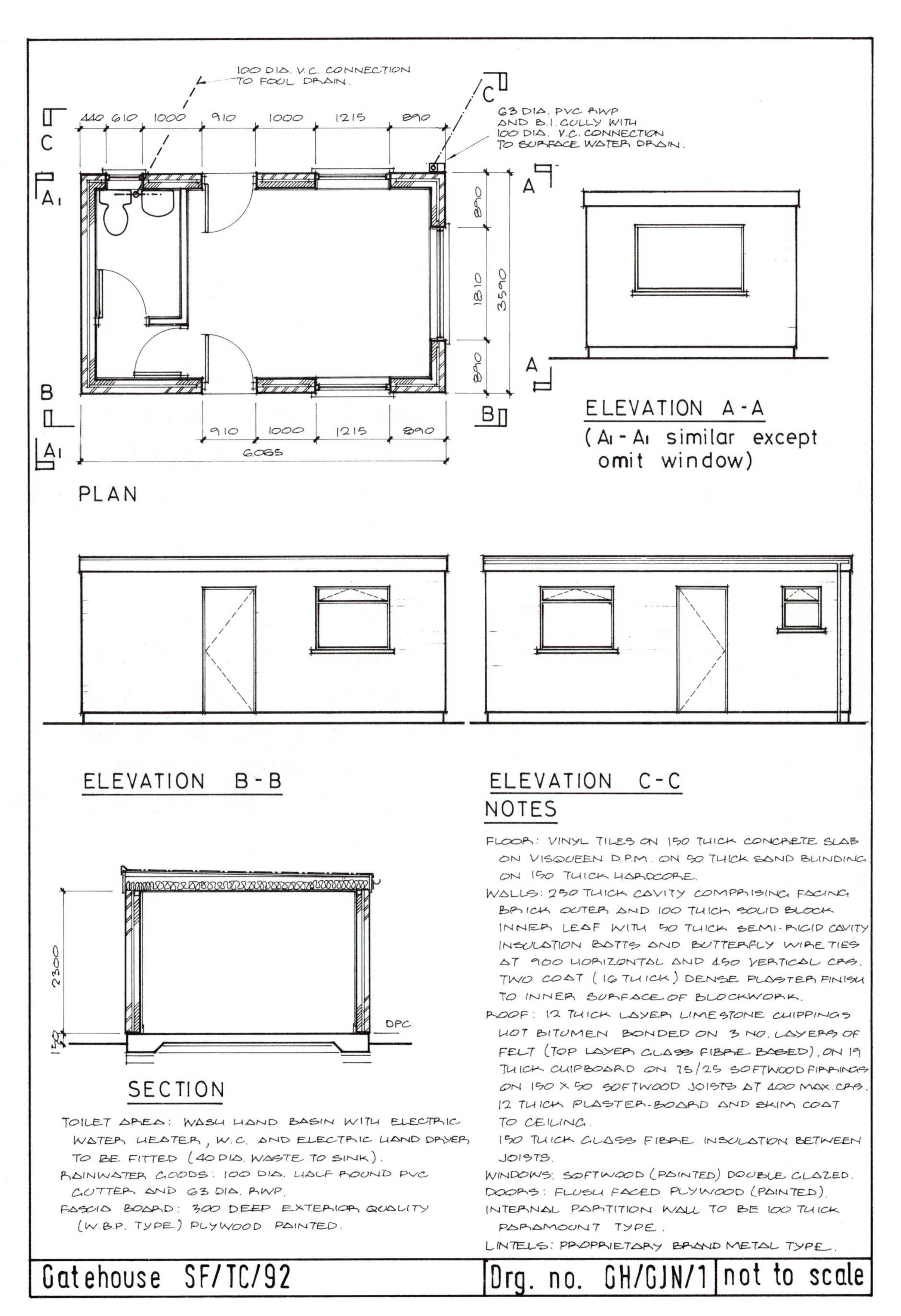
100 DIA. V.C. CONNECTION TO FOUL DRAIN.
63 DIA. PVC RWP AND B.I. GULLY WITH 100 DIA. V.C. CONNECTION TO SURFACE WATER DRAIN.
440 610 1000 910 1000 1215 890
890
1810
3590
890
910 1000 1215 890
6065
A
A
B
B
C
C
A1
A1
PLAN
ELEVATION A-A
(A1 - A1 similar except omit window)
ELEVATION B-B
ELEVATION C-C
2300
150
DPC
SECTION
TOILET AREA: WASH HAND BASIN WITH ELECTRIC WATER HEATER, W.C. AND ELECTRIC HAND DRYER TO BE FITTED (40 DIA. WASTE TO SINK).
RAINWATER GOODS: 100 DIA. HALF ROUND PVC GUTTER AND 63 DIA. RWP.
FASCIA BOARD: 300 DEEP EXTERIOR QUALITY (W.B.P. TYPE) PLYWOOD PAINTED.
NOTES
FLOOR: VINYL TILES ON 150 THICK CONCRETE SLAB ON VISQUEEN D.P.M. ON 50 THICK SAND BLINDING ON 150 THICK HARDCORE.
WALLS: 250 THICK CAVITY COMPRISING FACING BRICK OUTER AND 100 THICK SOLID BLOCK INNER LEAF WITH 50 THICK SEMI-RIGID CAVITY INSULATION BATTS AND BUTTERFLY WIRE TIES AT 900 HORIZONTAL AND 450 VERTICAL CRS. TWO COAT (16 THICK) DENSE PLASTER FINISH TO INNER SURFACE OF BLOCKWORK.
ROOF: 12 THICK LAYER LIMESTONE CHIPPINGS HOT BITUMEN BONDED ON 3 NO. LAYERS OF FELT (TOP LAYER GLASS FIBRE BASED), ON 19 THICK CHIPBOARD ON 75/25 SOFTWOOD FIRRINGS ON 150 X 50 SOFTWOOD JOISTS AT 400 MAX. CRS. 12 THICK PLASTER-BOARD AND SKIM COAT TO CEILING.
150 THICK GLASS FIBRE INSULATION BETWEEN JOISTS.
WINDOWS: SOFTWOOD (PAINTED) DOUBLE GLAZED.
DOORS: FLUSH FACED PLYWOOD (PAINTED).
INTERNAL PARTITION WALL TO BE 100 THICK PARAMOUNT TYPE.
LINTELS: PROPRIETARY BRAND METAL TYPE.
Gatehouse SF/TC/92
Drg. no. GH/GJN/1
not to scale

GATEHOUSE

Drawing No. GH/GJN/1

Explanatory Notes.

The following example shows a typical simple building which is often found on a water or sewage treatment works and is 'incidental to civil engineering works' and should be measured in accordance with the provisions of Class Z.

It is assumed that the foundations have been measured under Classes E, F and G, the brickwork under classes U and W and the decorating under Class V.

CARPENTRY AND JOINERY

Timesing	Dimensions	Squaring	Description	Notes
			Structural and carcassing timber; flat roofs.	
			6.07 overhang 2/0.75 .15 400 / 6.22 15.55 + 1 = 17	Divide the length of building by the joist centres to calculate the number of joists required.
			3.59 overhang 2/0.75 .15 3.74	
17/	3.74	63.58	Softwood joists; 150 x 50 mm (Z.113.1) & Firring pieces; 75 to 25 x 50mm (Z.113.2)	These tapered pieces (maximum height 75mm, minimum 25mm) create the slope to the roof.

			Sheet boarding; sloping upper surfaces.
	6.22		Chipboard 19mm
	3.74	23.26	thick.
			(z.132)
			Miscellaneous joinery.
2/	6.22	12.44	Fascia board; WPB
2/	3.74	7.48	exterior quality plywood;
		19.92	300x25mm
			(z.159)
			Skirtings;
2/	5.57	11.14	softwood.
2/	3.09	6.18	75 x 18mm
2/	1.40	2.80	(z.651) WC
	2.00	2.00	6.07
		22.12	wall thickness 2/250 .50
	Ddt.		5.57
4/	.91	3.64	doors 3.59
		18.48	2/.25 .50
			3.09

Insulation.

16/	3.74 .40	23.93	Glass fibre quilt 150mm thick, laid between joists) (Z.229)
			Windows, doors and glazing.
	1		Standard softwood window type 107V size 630x750mm (Z.311.1)
2/	1	2	Standard softwood window type 210W size 1200x1050mm (Z.311.2)
	1		Special observation window consisting of softwood members as specification clause 27.9; size 1810 x 1050mm. (Z.311.3)

	1	1	Standard flush plywood faced internal flush door, 40mm thick, size 1981 x 762mm (Z.313.1)
2/	1	2	Standard flush plywood faced external quality door, 44mm thick, size 1981x838 mm (Z.313.2)
	1	1	Standard softwood door lining size 27x94mm for door size 1981 x 762mm (Z.314.1)
2/	1	2	Standard softwood door frame size 33x64mm for door size 1981 x 838mm (Z.314.2)

Timesing	Dimension	Squaring	Description	Notes
2/	4.87	9.74	Miscellaneous joinery; softwood architraves once chamfered; size 75 x 18mm (Z.152)	
	4.95	4.95		
	4.95	4.95		
		19.64		
			Intl. door 2/1.98 3.96 2/.075 .15 .76 4.87	
			Extl doors 2/1.98 3.96 2/.075 .15 .84 4.95	
			<u>Ironmongery.</u>	
3/	1 pr	3 pr	Steel butt hinges; 100mm (Z.341)	Note deviation from CESMM. Hinges are usually bought and sold in pairs not enumerated. The change should be listed in the Preamble.
2/	1	2	Yale cylinder night rim latch (Z.343.1)	
	1	1	SAA mortice latch with lever handles (Z.343.2)	SAA is an abreviation for satin anodised aluminium.

	1	1	SAA wc indicating bolt (Z.343.3)	
2/	2	4	SAA Casement stay and fastener. (Z.349)	It is assumed that the windows are supplied without any ironmongery.
	1	1		
		5		
			<u>Glazing</u>	
2/	1.15 .70	1.61	Standard plain glass; clear float; to wood frames in putty; 6mm thick. (Z.351.1)	
2/	1.15 .18	.41		
	1.70 .95	1.61		
		3.94		
	.55 .40	.22	Standard plain glass; white patterned; to wood frames in putty 6mm thick (Z.351.2)	
	.55 .18	.09		
		.31		

Surface finishes, linings and partitions.

Timesing	Dimensions	Squaring	Description	Notes
	5.57 3.09	17.21	Insitu finishes, cement and sand (1:3) floor screed, steel trowelled finish, 25mm thick. (Z411) & Vinyl 'Polyflex' floor tiles, 2mm thick, fixed with adhesive (Z.421)	The screed and floor tiling has been measured under the partition although it may not be done this way on site.
	17.32 2.30	39.83	Insitu finishes, one coat cement and sand (1:3) backing coat, one coat Thistle class B plaster to blockwork, steel trowelled finish 16mm thick. (Z.413)	The plaster is measured overall and the doors and windows are then deducted. CESMM does not mention work to reveals so these areas are added in.
	17.55 .10	1.75 41.58		
Ddt.				
.63 .75	0.47			
1.20 1.05	2.52			
1.81 1.05	1.90			
2/0.91 2.00	3.64	8.53 32.95		

Waste calculations:

	2/5.57	11.14
	2/3.09	6.18
		17.32
Reveals Windows	2/.75	1.50
		.63
	2/2/1.05	4.20
	2/1.20	2.40
	2/1.05	2.10
		1.81
Doors.	2/2.00	4.00
		.91
		17.55

	5.57 3.09	17.21	Gypsum plasterboard fixing with nails to underside of softwood joists, 12.5mm thick (Z.434)
			&
			One coat Thistle board finish, steel trowelled, to soffit, 3mm thick. (Z.414)
	3.70 2.30	8.51	Paramount partition complete with softwood floor, wall and ceiling battens as specification clause 31.9, 57mm thick. (Z.479)
Ddt. .85 2.00		1.70 6.81	2.20 1.50 3.70
			&
			One coat Thistle board finish, steel trowelled to walls 3mm thick. (Z.413)

Piped building services.

			Pipework, copper pipe to BS2871, Table X lead free pre-soldered capillary joints, 15 mm diameter.	It is unusual for the pipework for the plumbing services to be shown on a drawing and the take off usually has to prepare an isometric sketch for his own use.
	1.00	1.00	to softwood skirting (Z.511.1)	
	1.00 1.75	2.75	to plastered wall (Z.511.2)	
			Extra over for	
	1	1	15 mm equal tee (Z.512.1)	
	4	4	15 mm elbow (Z.512.2)	
	1	1	tap connector (Z.512.3)	
	1	1	wc connector (Z.512.4)	
	1	1	water heater connector (Z.512.5)	

	1	1	Stopcock to BS1010, lead free pre-soldered capillary joints, gunmetal with brass headwork, 15mm diameter. (Z.512.6)
			Equipment.
	1	1	'Streamline' oversink water heater, 7 litre, 1kw, with spout and valve, fixed to plastered wall. (Z.529.1)
	1	1	'Handidry' electric hand drier, 1.4kw, fixed to wall. (Z.529.2)
			Sanitary appliances and fittings.
	1	1	Wash basin, vitreous china, complete with chromium plated waste, overflow with chain and plastic plug, polyprophylene P trap 40mm, 560x430mm white, wall mounted on brackets. (Z.530.1)

1	1	'Aztec' chromium plated pillar tap 15mm (Z.512.7)	
1	1	WC suite washdown type, vitreous china with plastic seat and cover, 9 litre cistern, ball valve, flush pipe, pan, connection to S trap. (Z.530.2)	
1.30	1.30	Ultra ABS marley waste system, solvent-welded joints, to clips, 40mm diameter. (Z.511.3)	
		Extra over for	
1	1	bend (Z.512.8)	

Drainage to structures above ground.

6.07	6.07	UPVC Terrain system rainwater gutter, straight half round, joint bracket joints, to softwood fascia with support brackets at 1m maximum centres (x.331)
1	1	Stop end (x.332.1)
1	1	Stop end outlet (x.332.2)
2.55	2.55	UPVC Terrain system rainwater pipe, straight, connector joints, to brickwork with clips at 2m centres, plugging (x.333)
1	1	Shoe (x.334)

Roofing.

			Three layers of bituminous felt roofing fibre based surface type 1B weighing 25 kg/10 m², limestone chipping to top surface.
	6.07 3.59	21.79	Upper surfaces inclined at an angle not exceeding 30° to the horizontal (W.341)
2/ 2/	6.07 3.59	12.14 7.18 19.32	Surfaces at width 100 mm (W.347)

Chapter 9
MENSURATION AND USEFUL DATA

The metric system

Linear

1 centimetre (cm)	=	10 millimetres (mm)
1 decimetre (dm)	=	10 centimetres (cm)
1 metre (m)	=	10 decimetres (dm)
1 kilometre (km)	=	1000 metres (m)

Capacity

1 millimetre (ml)	=	1 cubic centimetre (cm^3)
1 centilitre (cl)	=	10 millilitres (ml)
1 decilitre (dl)	=	10 centilitres (cl)
1 litre (l)	=	10 decilitres (dl)

Weight

1 centigram (cg)	=	10 milligrams (mg)
1 decigram (dg)	=	10 centigrams (cg)
1 gram (g)	=	10 decigrams (dg)
1 decagram (dag)	=	10 grams (g)
1 hectogram (hg)	=	10 decagrams (dag)
1 kilogram (kg)	=	10 hectograms (hg) = 1000 grams (g)

Imperial/metric conversions

Linear

1 in	=	25.4mm	1mm	=	0.03937in
1 ft	=	304.8mm	1cm	=	0.3937in
1 yd	=	914.4mm	1dm	=	3.937in
			1m	=	39.37in

Imperial/metric conversions (cont'd)

Square

1 in^2 = 645.16mm^2	1cm^2 = 0.155 in^2
1 ft^2 = 0.0929m^2	1m^2 = 10.7639 ft^2
1 yd^2 = 0.8361m^2	1m^2 = 1.196 yd^2

Cube

1 in^3 = 16.3871cm^3	1cm^3 = 0.061 in^3
1 ft^3 = 0.0283m^3	1m^3 = 35.3148 ft^3
1 yd^3 = 0.7646m^3	1m^3 = 1.307954 yd^3

Capacity

1 fl oz = 28.4ml	1ml = 0.0353 fl oz
1 pt = 0.568 ltr	1dl = 3.52 fl oz
1 gallon = 4.546 ltr	1 ltr = 1.7598 pt

Weight

1 oz = 28.35g	1g = 0.035 oz
1 lb = 0.4536kg	1kg = 35.274 oz
1 st = 6.35kg	1t = 2204.6 lb
1 ton = 1.016t	1t = 0.9842 ton

Temperature equivalents

In order to convert Fahrenheit to Celsius deduct 32 and multiply by 5/9. To convert Celsius to Fahrenheit multiply by 9/5 and add 32.

Fahrenheit		Celsius
230		110.0
220		104.4
212	Boiling point	100.0
210		98.9
200		93.3
190		87.8
180		82.2
170		76.7
160		71.1

Fahrenheit		Celsius
150		65.6
140		60.0
130		54.4
120		48.9
110		43.3
90		32.2
80		26.7
70		21.1
60		15.6
50		10.0
40		4.4
32	Freezing point	0.0
30		-1.1
20		-6.7
10		-12.2
0		-17.8

General information

Bricks Number of bricks m^2 in half brick thick wall in stretcher bond

50 x 102.5 x 215mm	74
65 x 102.5 x 215mm	59
75 x 102.5 x 215mm	52

Blocks Number of blocks per square metre

450 x 225mm	10
450 x 300mm	7
450 x 225mm	7

Timber 1 standard = 4.67227 cubic metres

1 cubic metre = 35.3148 cubic feet
10 cubic metres = 2.140 standards

Melting points of materials

Aluminium	658°C
Brass	927-1010°C
Bronze	912°C

Melting points of materials (cont'd)

Cast iron	1186°C
Copper	1083°C
Lead	327°C
Nickel	1452°C
Steel	1371°C
Tin	230°C
Zinc	419°C

Milled lead to BS1178

Code	Thickness	Weight	Colour code
3	1.32mm	14.97 kg/m^2	Green
4	1.80mm	20.41 kg/m^2	Blue
5	2.24mm	25.40 kg/m^2	Red
6	2.65mm	30.05 kg/m^2	Black
7	3.15mm	35.72 kg/m^2	White
8	3.55mm	40.26 kg/m^2	Orange

Soil properties

The study and classification of soils is clearly a subject of scientific investigation. The following information is provided as a guide only and should be used with caution. The figures given represent density in tonnes per m^3.

Soil	Loose	Compacted	Bearing capacity tonnes/m^2	Bulk* volume	Compacted* volume
Bog or peat	0.56	1.12	up to 2.0	1.1 - 1.3	-
Chalk	-	2.10	11 - 45	1.5 - 2.0	1.3 - 1.4
Clay - sandy	-	1.76	33 - 45	1.1 - 1.3	0.9 - 1.0
Clay - firm	-	1.92	45 - 65	1.3 - 1.4	0.9 - 1.0
Clay - stiff	-	2.08	65 -75	1.5	1.0
Gravel	1.76	1.92	66 - 90	1.0 - 1.30	0.9 - 1.0
Rock - soft	-	2.20	50 - 100	1.5 - 2.0	1.3 - 1.4
Rock - hard	-	2.70	100 - 200	1.5 - 2.0	1.3 - 1.4
Sand	1.44	1.76	22 - 40	1.0 - 1.10	0.9 - 1.0

* These figures are factors that will increase or decrease the net volume of undisturbed soil.

Soil definitions

SOFT - can be readily excavated with a spade and easily moulded with the fingers.

FIRM - can be excavated with a spade and moulded with substantial pressure by the fingers.

STIFF - requires a pick or pneumatic tool for excavation and cannot be moulded with the fingers.

Manholes - concrete surrounds

The following figures are net m^3 of concrete surround per metre of standard circular manhole segment to BS556.

nominal diameter	outside diameter	150mm surround	300mm surround
675mm	800mm	0.448	1.037
900mm	1048mm	0.565	1.271
1050mm	1219mm	0.645	1.432
1200mm	1397mm	0.729	1.599
1350mm	1575mm	0.813	1.767
1500mm	1727mm	0.885	1.910
1800mm	2032mm	1.028	2.198
2100mm	2388mm	1.196	2.533
2400mm	2692mm	1.339	2.820

Pipe beds and surrounds

The following are net quantities of material. Appropriate increases should be made for compaction, wastage and trench overbreak. All quantities are m^3 per metre of trench.

	Pipe diameter plus 300mm = trench width		Pipe diameter plus 600mm = trench width	
Pipe diameter (outside diameter)	150mm Bed	150mm Bed and surround	150mm Bed	150mm Bed and surround
100mm	0.060	0.152	0.105	0.272
150mm	0.068	0.185	0.113	0.320

Pipe beds and surrounds (cont'd)

Pipe diameter (outside diameter)	Pipe diameter plus 300mm = trench width: 150mm Bed	Pipe diameter plus 300mm = trench width: 150mm Bed and surround	Pipe diameter plus 600mm = trench width: 150mm Bed	Pipe diameter plus 600mm = trench width: 150mm Bed and surround
200mm	0.075	0.219	0.120	0.369
225mm	0.079	0.236	0.124	0.393
250mm	0.083	0.253	0.128	0.418
300mm	0.090	0.289	0.135	0.469
375mm	0.101	0.345	0.146	0.548
400mm	0.105	0.364	0.150	0.574
450mm	0.1125	0.403	0.158	0.628
500mm	0.120	0.444	0.165	0.684
600mm	0.135	0.527	0.180	0.797
750mm	0.158	0.661	0.203	0.976
900mm	0.180	0.804	0.225	1.164
1050mm	0.203	0.957	0.248	1.362
1200mm	0.225	1.119	0.270	1.569
1500mm	0.270	1.473	0.315	2.013

Velocity

To convert	Multiply by
Miles per hour into kilometres per hour	1.60934
Feet per second into metres per second	0.3048
Feet per minute into metres per second	0.00508
Feet per minute into metres per minute	0.30348
Inches per second into millimetres per second	25.4
Inches per minute into millimetres per second	0.42333
Inches per minute into centimetres per minute	2.54

Fuel consumption

To convert	Multiply by
Gallons per mile into litres per kilometre	2.825
Miles per gallon into kilometres per litre	0.354

Density

Tons per cubic yard into kilogrammes per cubic metre	1328.94
Pounds per cubic foot into kilogrammes per cubic metre	16.0185
Pounds per cubic inch into grammes per cubic centimetre	27.6799
Pounds per gallon into kilogrammes per litre	0.09978

Average plant outputs (cubic metres per hour)

Bucket size	Soil	Sand	Heavy clay	Soft rock
Face Shovel				
200	11	12	7	5
300	18	20	12	9
400	24	26	17	13
600	42	45	28	23
Backactor				
200	8	8	6	4
300	12	13	9	7
400	17	18	11	10
600	28	30	19	15
Dragline				
200	11	12	8	5
300	18	20	12	9
400	25	27	16	12
600	42	45	28	21

Bulkage of excavation

	Multiply volume by %
Soil	25
Gravel	15
Sand	12.5
Chalk	50
Clay (heavy)	30
Rock	30

Reinforcement mass

Hot rolled bars

Size in mm	Mass per metre in kg
6	0.222
8	0.395
10	0.616
12	0.888
16	1.579
20	2.466
25	3.854
32	6.313
40	9.864
50	15.413

Stainless steel bars

Size in mm	Mass per metre in kg
10	0.667
12	0.938
16	1.628
20	2.530
25	4.000
32	6.470

Mesh fabric

	Mesh size		Wire size		
	Main mm	Cross mm	Main mm	Cross mm	kg
A393	200	200	10	10	6.16
A252	200	200	8	8	3.95
A191	200	200	7	7	3.02
A142	200	200	6	6	2.22
A98	200	200	5	5	1.54
B1131	100	200	12	8	10.90
B785	100	200	10	8	8.14
B503	100	200	8	8	5.93
B385	100	200	7	7	4.53
B283	100	200	6	7	3.73
B196	100	200	5	7	3.05
C785	100	400	10	6	6.72
C503	100	400	8	5	4.34
C385	100	400	7	5	3.41
C283	100	400	6	5	2.61
D98	200	200	5	5	1.54
D49	100	100	2.5	2.5	0.77

Standard wire gauge and metric equivalent

SWG	mm	SWG	mm
3	6.40	15	1.83
4	5.89	16	1.63
5	5.38	17	1.42
6	4.88	18	1.21
7	4.47	19	1.02
8	4.06	20	0.91
9	3.63	21	0.81
10	3.25	22	0.71
11	2.95	23	0.61
12	2.65	24	0.56
13	2.34	25	0.51
14	2.03	26	0.46

Paper sizes

Size	mm	Inches
A0	841 x 1189	33.11 x 46.81
A1	594 x 841	23.39 x 33.11
A2	420 x 594	16.54 x 23.39
A3	297 x 420	11.69 x 16.54
A4	210 x 297	8.27 x 11.69
A5	148 x 210	5.83 x 8.27
A6	105 x 148	4.13 x 5.83
A7	74 x 105	2.91 x 4.13
A8	52 x 74	2.05 x 2.91
A9	37 x 52	1.46 x 2.05
A10	26 x 37	1.02 x 1.46

Weights of materials

Material	tonnes per m^3
Ashes	0.68
Aluminium	2.68
Asphalt	2.31
Brickwork - engineering	2.24
Brickwork - common	1.86
Bricks - engineering	2.40
Bricks - common	2.00

Weights of materials

Cement - Portland	1.45
Cement - rapid hardening	1.34
Clay - dry	1.05
Clay - wet	1.75
Coal	0.90
Concrete	2.30
Concrete - reinforced	2.40
Earth - topsoil	1.60
Glass	2.60
Granite - solid	2.70
Gravel	1.76
Iron	7.50
Lead	11.50
Limestone - crushed	1.75
Plaster	1.28
Sand	1.90
Slate	2.80
Tarmacadam	1.57
Timber - general construction	0.70
Water	1.00

INDEX